Topics in Analytic Number Theory

Topics in Analytic Number Theory

Edited by Sidney W. Graham and Jeffrey D. Vaaler

University of Texas Press, Austin

ISBN 978-0-292-74103-4
Library of Congress Catalog Card Number 85-50436

Printed in the United States of America

First Edition, 1985

For reasons of economy and speed this volume has been printed from camera-ready copy furnished by the editors, who assume full responsibility for its contents.

First paperback printing 2012

Contents

Preface

During the summer of 1982 a conference on analytic number theory was held at the University of Texas at Austin. The conference was supported by the University of Texas and by a grant from the National Science Foundation. There were three main speakers at this meeting, each of whom delivered a series of ten lectures. In addition, approximately twenty participants gave lectures on various parts of their own research. The main speakers were Professor Henryk Iwaniec of the Polish Academy of Sciences, Professor Hugh L. Montgomery of the University of Michigan, and Professor Heini Halberstam of the University of Illinois. Both Professor Iwaniec and Professor Halberstam wrote long papers on the subject of their lectures, which appear in this volume. Professor Montgomery's lectures covered material that will appear elsewhere. Eleven participants also contributed research or expository papers to the volume. As will be apparent from the table of contents, the topics covered by these proceedings represent a broad range of current research in analytic number theory. It is our hope that this collection of papers will make a useful contribution to research in this very active area of mathematics.

In preparing and editing the manuscripts for these proceedings we received help from many people. We especially wish to thank the University of Texas Press and Jan Duffy and Margaret Combs for their diligent and accurate typing.

S. W. Graham and J. D. Vaaler

A New Application of the Sieve to Probabilistic Number Theory

Krishnaswami Alladi

§1. INTRODUCTION, NOTATION AND BACKGROUND

Our object is to study the distribution of additive functions $f(n)$ for integers n belonging to subsets S of the positive integers $\mathbb{Z}^+$, via the estimation of their moments. Indeed there is a vast literature on such functions whose study forms the major portion of Probabilistic Number Theory; yet most results relate to the situation $S = \mathbb{Z}^+$ and little is known concerning distribution in subsets (see Elliot [5], Vols. 1 and 2). Although we do not solve the general problem, it was such an attempt that led us to a new method involving a crucial application of the sieve and the purpose of this paper is to describe this technique and mention a few applications.

For simplicity we concentrate here on strongly additive f, which are given by

$$f(n) = \sum_{p|n,\, p=\text{prime}} f(p) \quad ;$$

all results stated can be suitably modified to admit general additive functions which are only required to satisfy $f(mn) = f(m) + f(n)$, for $(m,n) = 1$. The sets S we consider will satisfy certain conditions imposed upon the quantity

$$S_d(x) = \sum_{\substack{n \leqslant x,\, n \in S \\ n \equiv 0 \pmod{d}}} a_n \quad ,$$

where $\{a_n\}$ are certain positive weights associated with S. For instance we require that

$$S_d(x) = \frac{X\omega(d)}{d} + R_d(x) \quad , \tag{1.1}$$

where $X = S_1(x)$, $\omega(d)$ is multiplicative, $0 \leqslant \omega(p) \leqslant 1$ and

"the average of $R_d(x)$ is in some sense small"; (1.2)

we shall make (1.2) more precise in the sequel.

Given a strongly additive f, and S as above, let

$$A(x) = A_f(x) = \sum_{p \leqslant x} \frac{f(p)\omega(p)}{p}$$

and

$$B_k(x) = \sum_{p \leqslant x} \frac{|f(p)|^k \omega(p)}{p}, \quad \text{for } k \geqslant 0 .$$

We will study the limiting behavior as $x \to \infty$ of the distribution function

$$F_x(v) = X^{-1} \sum_{\substack{n \leqslant x, n \in S \\ f(n) - A(x) < v\sqrt{B_2(x)}}} a_n$$

for $f \geqslant 0$. We shall do this by estimating the moments

$$\int_{-\infty}^{\infty} v^k dF_x(v) = \frac{1}{XB_2(x)^{k/2}} \sum_{\substack{n \leqslant x \\ n \in S}} a_n (f(n) - A(x))^k .$$

To achieve this we consider certain multiplicative functions g given in terms of the bilateral Laplace transform of F_x. In particular we look closely at sums involving certain of these g which satisfy $0 \leqslant g \leqslant 1$ and it is here that we use the sieve in a new way. It is the first occasion when the sieve has been used to estimate such moments and so we are able to treat a fairly wide class of sets S and additive functions f; our method may have other implications as well.

The first application of the sieve to the distribution of additive functions was by Erdös and Kac [7] who considered the values $f(n)$ for $1 \leqslant n \leqslant x$ in terms of nearly independent random variables and compared this with the distribution of certain sums of independent random variables. Here they needed Brun's sieve to formalize this comparison. These ideas have been developed since and in particular Kubilius [13] exploited this technique to the fullest by bringing in several powerful methods from Probability Theory. These methods however did not deal with moments.

It was Kac who first suggested that one ought to study the distribution of additive functions by estimating their moments. Halberstam [8; I,II,III] successfully carried out a portion of this project. His method was elementary and thus applied quite well to certain subsets of $\mathbb{Z}^+$; but it also involved many painstaking calculations. Subsequently this method was simplified and refined by Delange [3], [4], who considered suitable generating functions with product representations. So the moment method remained elementary and did not require sophisticated tools that were frequently used in other approaches.

By employing the sieve we retain in essense the advantages of Halberstam's approach, namely, applicability to subsets. The bilateral Laplace transform introduces some simplification and so some elegance of the Delange method is also retained. Our method stems from a recent technique due to Elliott [6], who neatly obtained uniform upper bounds for the absolute moments of arbitrary additive functions in the case $S = \mathbb{Z}^+$, $a_n \equiv 1$. The fusion of the sieve along with some of Elliott's ideas leads to this improved method.

Here we describe the main features of this method (sections 3, 4, and 5), and some results derived therefrom (§6). Only the main ideas behind a few proofs are briefly sketched (§7), and for full details see our recent paper [1]. In §8 we discuss a certain sieve theoretic aspect of the method and in §9 some limitations and merits are briefly mentioned. Some of Elliott's ideas are described in §2, but for a detailed comparitive study of our work against a classical background see [2].

All notation used so far will be retained. Always, in what follows, p will denote a prime, f a strongly additive function and g a strongly multiplicative function, namely, one that satisfies $g(n) = \prod_{p|n} g(p)$. Error estimates unless otherwise indicated are uniform or depend on S and this will be clear from the context.

§2. ELLIOT'S METHOD

In a recent paper Elliot [6] established that

$$\sum_{n\leqslant x} |f(n)-A(x)|^k \ll_k \begin{cases} xB_2(x)^{k/2} & \text{for } 0 \leqslant k \leqslant 2 \\ x(B_2(x)^{k/2} + B_k(x)) & \text{for } k > 2 \end{cases} \tag{2.1}$$

holds uniformly for all complex valued f and considered a few applications of (2.1) to the mean values of additive functions.

First he reduced the problem to that of proving (2.1) for $f \geqslant 0$ because we have the inequality

$$|a+b|^k \ll_k |a|^k + |b|^k \tag{2.2}$$

for all complex 'a' and 'b'. So if (2.1) holds for all $f\geqslant 0$, then its truth for all real f follows by writing real functions as the difference $f^+ - f^-$ of non-negative strongly additive functions generated by

$$f^+(p) = \max(0,f(p)) , \quad f^-(p) = -\min(0,f(p))$$

and using (2.2). Finally complex functions can be broken into their real and imaginary parts.

With this reduction he considers for $f \geqslant 0$ and complex z, the sum

$$T_z(x) = \sum_{n \leqslant x} e^{z(f(n) - A(x))}$$

and notes that for even $k \geqslant 0$ the left side of (2.1) is

$$\frac{k!}{2\pi i} \int_{|z|=r} \frac{T_z(x)dz}{z^{k+1}} \leqslant \frac{k!}{2\pi} \int_{|z|=r} \frac{T_u(x)|dz|}{r^{k+1}}, \qquad (2.3)$$

where $u = \mathrm{Re}(z)$. If (2.1) can be established for even $k \geqslant 0$ using (2.3), then the truth of (2.1) for all $k \geqslant 0$ follows by a suitable application of the Hölder-Minkowski inequality. We rewrite

$$T_u(x) = e^{-uA(x)} \sum_{n \leqslant x} e^{uf(n)} = e^{-uA(x)} \sum_{n \leqslant x} g(n)$$

and note that g satisfies either

<u>Case 1</u> $\quad u \leqslant 0 \Rightarrow 0 < g(n) \leqslant 1$ for all n, or

<u>Case 2</u> $\quad u > 0 \Rightarrow g(n) \geqslant 1$ for all n.

Elliot takes care of Case 1 by appealing to an upper bound due to Hall [10], for sums of submultiplicative functions, i.e. : those $\tilde{g}$ satisfying $0 \leqslant \tilde{g}(mn) \leqslant \tilde{g}(m)\tilde{g}(n)$. With regard to Case 2 he defines h such that $\sum_{d|n} h(d) = g(n)$. This gives h to be multiplicative, $\geqslant 0$, and $h(p^e) = 0$ for all p, and $e \geqslant 2$. Therefore

$$\sum_{n \leqslant x} g(n) \leqslant x \sum_{d \leqslant x} \frac{h(d)}{d} \leqslant x \prod_{p \leqslant x} \left(1 + \frac{h(p)}{p}\right). \qquad (2.4)$$

These estimates are sufficient to establish (2.1) when the $f(p)$ do not take "large values". Large here means those p for which $f(p)/\sqrt{B_2(x)}$ is large. Such primes are "sparse" because

$$\sum_{\substack{p\leqslant x\\ f(p)\geqslant\delta\sqrt{B_2(x)}}} \frac{1}{p} \leqslant \frac{1}{\delta^2} \sum_{p\leqslant x} \frac{f^2(p)}{pB_2(x)} = \delta^{-2} . \tag{2.5}$$

So he could control the contribution due to these sporadic large values in a manner described in §7. We do not give further details here.

Our goal is to estimate the moments asymptotically so that we may determine the weak limit of $F_x(v)$ in certain cases. This means we have to do away with a simple inequality like (2.3) and use a superior inversion technique. A lemma of this nature is stated in the next section.

§3. TWO TAUBERIAN LEMMAS

<u>Lemma 1</u>: Let $\varphi_x(v)$ be a sequence of probability distributions and $R > 0$ such that

$$\int_{-\infty}^{\infty} e^{uv} dF_x(v) \ll 1 \quad \text{for} \quad -R \leqslant u \leqslant R . \tag{3.1}$$

Then

$$\int_{-\infty}^{\infty} |v|^k d\varphi_x(v) \ll \frac{k!}{R^k} , \quad \text{for} \quad k = 1,2,3\ldots .$$

If in addition to (3.1) we have

$$\lim_{x\to\infty} \int_{-\infty}^{\infty} e^{uv} d\varphi_x(v) = \ell(u) , \quad u \leqslant 0 \tag{3.2}$$

exist and is finite, with uniform convergence in $-R \leqslant u \leqslant 0$, then

$$\lim_{x\to\infty} \int_{-\infty}^{\infty} v^k d\varphi_x(v) = m_k$$

exists and is finite for $k = 1,2,3,\dots$.

If (3.1) and (3.2) hold, then there is a probability distribution $\varphi(v)$ such that

$$m_k = \int_{-\infty}^{\infty} v^k d\varphi(v), \quad \text{for } k = 1,2,3,\dots$$

and

$$\varphi_x(v) \to \varphi(v) \text{ weakly as } x \to \infty \quad .$$

Finally if $\ell(u)$ can be extended analytically into $|z| \leqslant R$, then

$$m_k = \frac{d^k}{dz^k} \ell(z)\Big|_{z=0} \quad \text{for } k = 1,2,3,\dots \; . \quad //$$

For a proof of this lemma see [1].

We shall apply the lemma with $\varphi_x(v)$ taken to be equal to

$$F_{x,y}(v) = \frac{1}{X} \sum_{\substack{n\leqslant x, n\in S \\ f_y(n)-A(y) < v\sqrt{B_2(y)}}} a_n \quad , \tag{3.3}$$

where $f \geqslant 0$ and

$$f_y(n) = \sum_{p|n, p\leqslant y} f(p) \quad .$$

We shall choose y optimally such that

$$\alpha = \log x/\log y \to \infty \text{ with } x \; .$$

With $F_{x,y}(v)$ as above we have

$$\int_{-\infty}^{\infty} e^{uv} dF_{x,y}(v) = \frac{T_u(x,y)}{X} = \frac{e^{-uA(y)/\sqrt{B_2(y)}}}{X} \sum_{\substack{n\leqslant x \\ n\in S}} e^{uf_y(n)/\sqrt{B_2(y)}} \tag{3.4}$$

Condition (3.1) of Lemma 1 implies that we now need an asymptotic estimate for $T_y(x,y)$ when $u \leqslant 0$ (Case 1). This we achieve using a sieve method (see §4). With regard to $u > 0$ (Case 2), condition (3.2) of Lemma 1 tells us that a good upper bound for $T_u(x,y)$ suffices. Here the idea underlying (2.4) has to be developed and there will be more on this in §5.

There are instances when one can derive asymptotic estimates for moments after first deriving good upper bounds for them. This is the idea behind

Lemma 2: Let $\varphi_x(v)$ be a sequence of probability distribution and $\varphi(v)$ a probability distribution such that $\varphi_x(v) \to \varphi(v)$ weakly in v as $x \to \infty$. In addition assume that

$$\int_{-\infty}^{\infty} v^k d\varphi_x(v) \ll_k 1 \quad \text{for all } x . \tag{3.5}$$

Then

$$\lim_{x\to\infty} \int_{-\infty}^{\infty} v^k d\varphi_x(v) = \int_{-\infty}^{\infty} v^k d\varphi(v) \tag{3.6}$$

holds for $k = 1,2,3,\ldots,$ where the right hand side of (3.6) is finite. //

The proof of Lemma 2 is similar to that of a weaker version given in [1]. Lemma 2 is crucial to Theorem 2 which deals with distribution in the set

$$S^{(1)} = \{p+a \mid p = \text{prime}\}, \quad \text{where } a \in \mathbb{Z}^+ . \tag{3.7}$$

The principle underlying Lemma 2 may have other useful applications.

§4. A NEW APPLICATION OF THE COMBINATORIAL SIEVE

A standard problem in Sieve Theory is the estimation of $S(x,y)$, the sum of the weights a_n, for $n \in S$ which are

$\leqslant x$ and devoid of prime factors $\leqslant y$. Let $P_y = \prod_{p \leqslant y} p$ and μ the Möebius function. Therefore

$$S(x,y) = \sum_{\substack{n \leqslant x \\ n \in S, (n,P_y)=1}} a_n = \sum_{\substack{n \leqslant x \\ n \in S}} a_n \sum_{d | (n,P_y)} \mu(d) \tag{4.1}$$

A simple rearrangement of (4.1) using (1.1) yields

$$S(x,y) = \sum_{d | P_y} \mu(d) S_d(x) = X \sum_{d | P_y} \frac{\mu(d)\omega(d)}{d} + O\Big(\sum_{d | P_y} |R_d(x)| \Big) . \tag{4.2}$$

When y is large the divisors of P_y are too numerous and so the error term in (4.2) becomes unwieldly.

The combinatorial sieve overcomes this difficulty in the following manner. We take functions χ^- and χ^+ satisfying

$$\chi^-(1) = \chi^+(1) = 1 \tag{4.3}$$

and

$$\sum_{d|n} \mu(d)\chi^-(d) \leqslant \sum_{d|n} \mu(d) \leqslant \sum_{d|n} \mu(d)\chi^+(d) \quad \text{for all} \quad n \in \mathbb{Z}^+ . \tag{4.4}$$

The sum involving χ^+ in (4.4) when substituted suitably in (4.1) yields an upper bound for $S(x,y)$, whereas the sum involving χ^- supplies a lower bound. We also choose χ^-, χ^+ to vanish often enough on the divisors of P_y to keep the error term under control. If the upper and lower bounds obtained by such choices are close, we then have an asymptotic estimate for $S(x,y)$.

Consider a set S for which the R_d satisfy the following condition: there is a number β with $0 < \beta \leqslant 1$ such that for each $b > 0$ there is a $c > 0$ with the property

$$\sum_{d \leqslant x^{\beta}/(\log X)^{c}} |R_d(x)| \ll_b \frac{X}{\log^b X} . \tag{4.5}$$

In addition assume there is $c > 0$ such that $|R_d(x)| \ll (X \log X/d + 1)c^{\nu(d)}$, where $\nu(d) = \sum_{p|d} 1$. For such a set it is known by the combinatorial sieve that

$$S(x,y) \sim X \prod_{p \leqslant y} \left(1 - \frac{\omega(p)}{p}\right), \quad \text{as} \quad x, \alpha \to \infty \tag{4.6}$$

(see Halberstam and Richert [9], p. 83).

We shall now employ the combinatorial sieve to estimate

$$\sum_{\substack{n \leqslant x \\ n \in S}} a_n g(n) , \quad \text{when} \quad 0 \leqslant g \leqslant 1 . \tag{4.7}$$

This is made possible by the following:

Lemma 3: Let χ^-, χ^+ satisfy (4.3) and (4.4). Let g^* be a multiplicative function satisfying $0 \leqslant g^* \leqslant 1$. Then for all $n \in \mathbb{Z}^+$ we have

$$\sum_{d|n} \mu(d) g^*(d) \chi^-(d) \leqslant \sum_{d|n} \mu(d) g^*(d) \leqslant \sum_{d|n} \mu(d) g^*(d) \chi^+(d) . \quad //$$

Lemma 3 follows easily from (4.3),(4.4) and the Möebius Inversion formula and a proof can be found in [1].

Let g be as in (4.7). We define g^* to be the function satisfying

$$g(n) = \sum_{d|n} \mu(d) g^*(d) . \tag{4.8}$$

Therefore g^* is multiplicative and $0 \leqslant g^* \leqslant 1$. Furthermore

$$g_y(n) = \prod_{\substack{p|n \\ p\leqslant y}} g(p) = \sum_{d|(n,P_y)} \mu(d)g^*(d) .$$

Therefore

$$\sum_{\substack{n\leqslant x \\ n\in S}} a_n g_y(n) = \sum_{\substack{n\leqslant x \\ n\in S}} a_n \sum_{d|(n,P_y)} \mu(d)g^*(d) . \qquad (4.9)$$

The sums involving χ^- and χ^+ of Lemma 3, when substituted into (4.9) yield upper and lower bounds. If we work out the details as in the combinatorial sieve the main term that we get for (4.9) is

$$X \prod_{p\leqslant y} \left(1 - \frac{\omega(p)g^*(p)}{p}\right) ,$$

which is at least as large as the right hand side of (4.6). On the other hand the principal error term in this calculation is

$$\sum_{d \text{ sq. free}\leqslant x} |R_d(x)|g^*(d)|\chi^{\pm}(d)|$$

which is

$$\leqslant \sum_{d \text{ sq. free}\leqslant x} |R_d(x)||\chi^{\pm}(d)| ,$$

the principal error term in the combinatorial sieve. So the situation is more favourable for us and we get the following result

Lemma 4: Let g be strongly multiplicative, $0 \leqslant g \leqslant 1$, and g^* as in (4.8). Let $\alpha \to \infty$, and S satisfy (1.1), and (4.5). Then

$$\sum_{\substack{n \leqslant x \\ n \in S}} a_n g_y(n) \sim X \prod_{p \leqslant y} \left(1 - \frac{\omega(p) g^*(p)}{p}\right) \quad \text{as} \quad x \to \infty, \tag{4.10}$$

where the estimate is uniform for all such g (i.e., the error term depends only on S, x and α). Also, since $g_y(n) \leqslant g(n)$, we have uniformly

$$\sum_{\substack{n \leqslant x \\ n \in S}} a_n g(n) \ll X \prod_{p \leqslant x} \left(1 - \frac{\omega(p) g^*(p)}{p}\right). \quad //$$

§5. AUXILIARY UPPER BOUNDS

In addition to the estimates of Lemma 4 we require bounds for

$$\sum_{\substack{n \leqslant x \\ n \in S}} a_n g(n) \text{ , when } g \geqslant 1 \text{ .} \tag{5.1}$$

Using the function h employed by Elliott (see (2.4)) we see that the sum in (5.1) is

$$X \sum_{d \leqslant x} \frac{h(d)\omega(d)}{d} + \sum_{d \leqslant x} h(d) R_d(x) \text{ .} \tag{5.2}$$

The second term in (5.2) is complicated to handle unless the R_d are quite well behaved. In Elliott's case we had $|R_d(x)| \leqslant 1$.

Suppose that

$$|R_d(x)| \ll \frac{X\omega(d)}{d} \text{ , } \quad \text{for} \quad 1 \leqslant d \leqslant x \text{ ;} \tag{5.3}$$

then from (5.1) and (5.3) we deduce that the sum in (5.1) is

$$\ll X \prod_{p \leqslant x} \left(1 + \frac{\omega(p) h(p)}{p}\right) \text{ .} \tag{5.4}$$

Examples of sets satisfying (5.3) are

(i) $\mathbb{Z}^+$ or an arithmetic progression of positive integers with $a_n \equiv 1$.

(ii) The integers n having $\nu(n) \equiv 1 \pmod 2$ (or $\nu(n) \equiv 0 \pmod 2$) with $a_n \equiv 1$.

(iii) The integers $n \leqslant x$ which can be written as a sum of two integer squares, with $a_n = r(n)$, the number of such representations.

There are however many interesting sets where the inequality (5.3) simply does not hold in such a long range of values d, but remains true on a shorter interval. For instance if

$$S = S^{(2)} = \{Q(n) \mid n = 1,2,3,\ldots\}, \quad \text{where} \quad Q(x) \in \mathbb{Z}^+[x] \ ,$$

then (5.3) holds for $d \leqslant x^{1/\beta}$, with $\beta = \deg Q$.

Similarly if $S = S^{(1)}$ (see (.7)), then (5.3) holds (with $\ll_\beta$) for all $d \leqslant x^{1/\beta}$, where $0 < \beta < 1$. In particular (5.3) holds for $d \leqslant \sqrt{x}$. This follows from the Brun-Titchmarch inequality, well known in Sieve Theory (see [9], Ch. 3). In such situations the following lemma is useful.

Lemma 5: Let h be multiplicative and $0 \leqslant h \leqslant 1$. Then for all square-free integers n we have

$$\sum_{\substack{d \mid n \\ d \leqslant \sqrt{n}}} h(d) \geqslant \sum_{\substack{d \mid n \\ d > \sqrt{n}}} h(d) \ . \quad //$$

There are many ways of proving Lemma 5 and we briefly discuss the ideas behind two proofs.

Recently I conjectured that for each square-free n, there is a one-to-one mapping m, between the divisors d of n that are $\leqslant \sqrt{n}$ and those divisors d' of n which are $\geqslant \sqrt{n}$ such that

$$m(d) = d' \equiv 0 \pmod{d} . \tag{5.5}$$

Very recently (in fact during this conference!) Erdös and Vaaler showed by induction on $\nu(n)$, that more is true. They proved that for each $t \leqslant \sqrt{n}$, there is a one-to-one mapping m_t between the divisors d of n which are $\leqslant t$ and those divisors d' of n which are $\geqslant n/t$ such that

$$m_t(d) = d' \equiv 0 \pmod{d} . \tag{5.6}$$

Thus (5.6) implies (5.5), which gives Lemma 5.

Motivated by certain distribution properties of the divisor function $\tau(n)$ in short intervals, Hall and Tenenbaum [11] have shown by induction on $\nu(n)$, that if H is a non-negative multiplicative function such that for each prime p, the sequence $\{H(p^e)\}$, $e = 0,1,2,3,\ldots$ is non-increasing, then

$$\frac{\sum_{d|n, d\leqslant t} H(d)}{\sum_{d|n} H(d)} \geqslant \frac{\sum_{d|n, d\leqslant t} 1}{\tau(n)} , \tag{5.7}$$

holds for all n, and all $t \leqslant \sqrt{n}$. Thus (5.7) implies Lemma 5.

What makes these two proofs work is that the induction hypothesis is very strong, namely, the truth of the statements for all $t \leqslant \sqrt{n}$. Erdös remarked that one reason I could not prove (5.5) was because it was a weaker conjecture -- my induction hypothesis was not strong enough!

These two proofs of Lemma 5 along with one more due to Heath-Brown (private communication) will be discussed in greater detail in a subsequent paper where we will consider separately the distribution of additive functions in $S^{(1)}$. Hall has also informed us that Woodall a few years ago arrived at our mapping (5.5) and had proved it combinatorially. At any rate this is the first occasion when Lemma 5 has been applied to the distribution of additive functions.

The restriction $h \leqslant 1$ (equivalently $g(p) \leqslant 2$) is not very severe. In fact in many of our applications this suffices. Clearly Lemma 5 applies to $S^{(1)}$, and to $S^{(2)}$ provided $\deg Q \leqslant 2$. When $\deg Q > 2$ certain problems arise. The R_d here satisfy conditions much weaker than (5.3) and so as of now we are required to impose more severe growth conditions upon h, and therefore on the underlying f as well. We shall not delve into this further right now. However, an inequality such as (5.7) might prove quite useful in such situations, and this may be worth pursuing further.

When g is strongly multiplicative and $1 \leqslant g(p) \leqslant 2$ we have from Lemma 5

$$g(n) = \sum_{d|n} h(d) \leqslant 2 \sum_{\substack{d|n \\ d \leqslant \sqrt{n}}} h(d) \quad .$$

Therefore we arrive at the following result.

<u>Lemma 6</u>: Let $g \geqslant 1$. If S satisfies (5.3), then (5.4) holds uniformly for all such g. If S satisfies the weaker condition

$$|R_d(x)| \ll \frac{X\omega(d)}{d}, \quad \text{for} \quad 1 \leqslant d \leqslant \sqrt{x} \quad , \tag{5.8}$$

then (5.4) holds uniformly for all such g satisfying $1 \leqslant g(p) \leqslant 2$. //

§6. SOME RESULTS

For S satisfying (1.1) and (4.5) we consider the collection $\mathcal{A}(S)$ of all f satisfying the following two conditions

(i) There exists some $\alpha = \alpha(x) \to \infty$ with x such that

$$\lim_{x\to\infty} \{B_2(x)/B_2(y)\} = 1 .$$

(ii) For all x we have

$$\{\max_{p\leqslant x} f(p)\}/\sqrt{B_2(x)} \leqslant 1 .$$

From (i) and (ii) we deduce that there is $\alpha \to \infty$ such that

$$\lim_{x\to\infty} \frac{B_k(x)-B_k(y)}{B_2(x)^{k/2}} = 0 \quad \text{for } k = 1,2,3,\ldots . \tag{6.1}$$

For this class $\mathcal{a}$ we can establish the following distribution theorems via the asymptotic estimation of moments.

Theorem 1: Let S satisfy (1.1), (4.5) and (5.3). Let $0 \leqslant f \in \mathcal{a}$ and suppose that $B_2(x) \to \infty$. Assume also that there is a probability distribution K such that

$$\lim_{x\to\infty} \frac{1}{B_2(x)} \sum_{\substack{p\leqslant x \\ f(p)\leqslant u\sqrt{B_2(x)}}} \frac{f^2(p)\omega(p)}{p} = K(u)$$

almost surely in u. Then

$$\lim_{x\to\infty} \frac{1}{XB_2(x)^{k/2}} \sum_{\substack{n\leqslant x \\ n\in S}} (f(n)-A(x))^k = M_k \tag{6.2}$$

exists and is finite for $k = 1,2,3,\ldots$.

In addition there is a probability distribution $F(v)$ such that

$$\int_{-\infty}^{\infty} v^k dF(v) = M_k , \quad \text{for } k = 1,2,3,\ldots \tag{6.3}$$

and

$$F_x(v) \to F(v) \quad \text{weakly in} \quad v \quad \text{as} \quad x \to \infty . \tag{6.4}$$

Furthermore the function

$$L(z) = \exp\left\{\int_{-\infty}^{\infty} \frac{e^{zv}-1-zv}{v^2}\, dK(v)\right\} \tag{6.5}$$

is analytic in a circle $|z| \leqslant R$ about the origin with $L(iu)$ as the characteristic function of F. Therefore

$$M_k = \left.\frac{d^k L(z)}{dz^k}\right|_{z=0} \quad \text{for} \quad k = 1,2,3,\ldots \; . \; // \tag{6.6}$$

Theorem 2: (a) Let $S = S^{(1)}$ as in (3.7). Let $f \in \mathcal{A}(S^{(1)})$ satisfy the hypothesis of Theorem 1. Then (6.4) holds with $L(z)$ in (6.5) as the characteristic function of F.

(b) Moreover (6.2) and (6.3) hold as well with M_k being given by (6.6). //

Theorem 1 can be established under slightly weaker growth conditions upon f (see [1]). Barban (see [5], Vol. 2.p.27) using a method due to Kubilius and a trick (see §7) established Theorem 2(a). However, it was not known whether under that hypothesis statements (6.2) and (6.3) also hold for $f \in \mathcal{A}(S^{(1)})$ and part (b) now settles this question. In the sequel we shall use Lemma 2 to deduce part (b) from part (a).

There are versions of the above theorems for the case $B_2(x) \to B < \infty$.

Theorem 3: Let S satisfy the hypothesis of Theorem 1. Let $0 \leqslant f \in \mathcal{A}(S)$ satisfy $B_2(x) \to B < \infty$. Then

$$\lim_{x\to\infty} \frac{1}{X} \sum_{\substack{n\leqslant x\\ n\in S}} a_n(f(n)-A(x))^k = M_k^*$$

exists and is finite for $k = 1,2,3,\ldots$.

In addition there is a probability distribution F^* such that

$$M_k^* = \int_{-\infty}^{\infty} v^k dF^*(v) \ , \quad \text{for} \quad k = 1,2,3,\ldots$$

and

$$\frac{1}{X} \sum_{\substack{n\leqslant x, n\in S\\ f(n)-A(x)<v}} a_n \to F^*(v) \quad \text{weakly as} \quad x\to\infty .$$

Finally

$$L^*(z) = \prod_p \left(1 + \frac{(e^{zf(p)}-1)\omega(p)}{p}\right) e^{-zf(p)\omega(p)/p}$$

is analytic in a circle $|z| \leqslant R^*$ about the origin with $L^*(iu)$ as the characteristic function of F^*. Therefore

$$M_k^* = \left.\frac{d^k L^*(z)}{dz^k}\right|_{z=0} \qquad \text{for} \quad k = 1,2,3,\ldots . \quad //$$

An analogue of Theorem 3 holds for $S^{(1)}$ but we will not state it here.

Previously in [1] I had observed Lemma 2 only when φ was continuous but Professor Elliott showed me recently that the continuity of φ is unnecessary. Consequently that assumption is unnecessary now in Theorem 2(b) and in Theorem 3 for $S^{(1)}$.

If f attains very large values on certain p then we cannot estimate the moments asymptotically. However, it is possible to derive good upper bounds, like those of Elliott, in such cases. More generally we have

Theorem 4: Let S satisfy the hypothesis of Theorem 1. Then for all f we have

$$\sum_{\substack{n\leqslant x\\ n\in S}} a_n |f(n)-A(x)|^k \ll_k \begin{cases} XB_2(x)^{k/2} & \text{for } 0\leqslant k\leqslant 2 \\ X(B_2(x)^{k/2}+B_k(x)) & \text{for } k>2\ .// \end{cases}$$

Although we don't give details, we shall, in the next section, describe briefly the main ideas behind the proofs of these theorems.

§7. SKETCH OF PROOFS

We being with Theorem 1.

First we choose $y,\alpha\to\infty$ with x such that (6.1) holds. From (3.4), and the asymptotic estimate (4.10) of Lemma 4 we see after computation of $T_u(x,y)$ that (3.2) in Lemma 1 holds with $\varphi_x(v)$ set equal to $F_{x,y}(v)$. Similarly from Lemma 6 we see that (3.1) holds for $u\geqslant 0$, for $F_{x,y}(v)$ and the truth of this statement for $u<0$ trivially follows from (3.2). The radius R is given in terms of the implicit constant in (ii). Also $\ell(z)$ is identified with $L(z)$ and the analyticity property in $|z|\leqslant R$ follows from (ii). So Theorem 1 has been proved for $f_y(n)$ in the place of $f(n)$.

Once this is done we deduce the truth of Theorem 1 for $f(n)$ by appealing to

$$\sum_{n\leqslant x, n\in S} a_n(f(n)-f_y(n))^k = o(XB_2(x)^{k/2})\ . \qquad (7.1)$$

This is a consequence of (i) and (ii) along with a suitable application of the Holder-Minkowski inequality. For details see [1].

While proving Theorem 2 the first part of the above proof remains unchanged; namely, treatment of Cases 1 and 2 by means of Lemmas 4 and 6, which establishes asymptotic estimates for the moments of $f_y(n)$ with $n \in S^{(1)}$. However it is a bound such as (7.1) which gives trouble because here we do not have (5.3) for $d \leqslant x$. Instead we have (5.8) and therefore only

$$\sum_{\substack{n \leqslant x \\ n \in S^{(1)}}} a_n (f_Y(n) - f_y(n))^k = o(XB_2(x)^{k/2}) , \tag{7.2}$$

where

$$\log X / \log Y = 2k . \tag{7.3}$$

This establishes (6.2) for $f_Y(n)$.

Observe that (7.3) and (ii) imply

$$f(n) - f_Y(n) \ll 2k\sqrt{B_2(x)}$$

and therefore from (2.2),(7.2) and Theorem 2 for $f_Y(n)$ we get

$$\sum_{\substack{n \leqslant x \\ n \in S^{(1)}}} a_n |f(n) - A(x)|^k \ll_k XB_2(x)^{k/2} . \tag{7.4}$$

At this point the following lemma due to Barban is useful (see [5], Vol. 1, p. 173 for a proof).

<u>Lemma 7</u>: Let f^* be an arbitrary strongly additive function and $0 < \beta < 2$. Then there exists $D(\beta)$ such that

$$\sum_{\substack{n \leqslant x \\ n \in S^{(1)}}} |f^*(n) - A^*(x)|^\beta \ll D(\beta) \cdot \frac{x}{\log x} B_2^*(x)^{\beta/2} ,$$

where $A^*(x) = A_{f^*}(x)$ and B_2^* is defined similarly. //

We know that Theorem 2 holds for f_y in the place of f and so we need to only fill the gap. We set $f^* = f - f_y$ in

Lemma 7 and note that (6.1) gives

$$A^*(x) = o(\sqrt{B_2(x)}) \quad \text{and} \quad B_2^*(x) = o(B_2(x)) \quad .$$

Therefore

$$\text{(weakly)} \lim_{x\to\infty} F_x(v) = \lim_{x\to\infty} F_{x,y}(v) = F(v) \qquad (7.5)$$

This proves Theorem 2, part (a). Part (b) follows from (a), (7.5) and Lemma 2 when (7.4) is identified with (3.5).

Lemma 2 has other uses also, one of which is the extension of Theorems 1,2, and 3 to real valued $f \in \mathcal{A}$ and this is indicated in §9.

Theorem 3 is proved in a manner similar to Theorem 1. Here the hypothesis (i) is redundant because $B_2(x) \to B < \infty$. So the details are simpler.

With regard to Theorem 4, as Elliott has shown us, it suffices to prove it for $f \geqslant 0$. We decompose f into $f_1 + f_2$ which are given by

$$f_1(p) = \begin{cases} f(p) & \text{if} \quad f(p) \leqslant \sqrt{B_2(x)} \\ 0 & \text{if} \quad f(p) > \sqrt{B_2(x)} \end{cases}$$

and

$$f_2(p) = \begin{cases} 0 & \text{if} \quad f(p) \leqslant \sqrt{B_2(x)} \\ f(p) & \text{if} \quad f(p) > \sqrt{B_2(x)} \end{cases}$$

Our treatment of Cases 1 and 2 in Theorem 1 shows that

$$\sum_{\substack{n\leqslant x \\ n\in S}} a_n |f_1(n)-A_{f_1}(x)|^k \ll_k XB_2(x)^{k/2} \quad . \qquad (7.6)$$

With regard to f_2 we have analogous to (2.5)

$$\sum_{\substack{p \leqslant x \\ f_2(p) > 0}} \frac{\omega(p)}{p} \leqslant 1 . \tag{7.7}$$

It is not difficult to see that (7.7) gives

$$\sum_{\substack{n \leqslant x \\ n \in S}} a_n |f_2(n) - A_{f_2}(x)|^k \ll_k XB_2(x) \quad . \tag{7.8}$$

Theorem 4 follows from (7.6), (7.8), and (2.2).

§8. SIEVE DIMENSION

In many applications of the sieve method one does not sieve a set S by the set $\mathcal{P}^{(\infty)}$ of all primes. Instead one has a finite or infinite set $\mathcal{P}$ of primes and requires an estimate for

$$S(x,y,\mathcal{P}) = \sum_{\substack{n \leqslant x, n \in S \\ p|n, p \leqslant y \Rightarrow p \notin \mathcal{P}}} a_n \tag{8.1}$$

The dimension κ of the sieve process in (8.1) is a number that satisfies

$$\prod_{\substack{2 \leqslant y_1 \leqslant p \leqslant y_2 \\ p \in \mathcal{P}}} \left(1 - \frac{\omega(p)}{p}\right)^{-1} \ll \left(\frac{\log y_2}{\log y_1}\right)^{\kappa} . \tag{8.2}$$

Clearly from this definition we see that κ is not unique and that it pays to choose the smallest κ. There are many results where the size of the error terms depend on κ and decrease in size with κ. Thus, as a general principle, one can deduce from a result for a sieve of dimension κ, a corresponding statement of at least the same quality for a related sieve problem of smaller dimension. For sets S which have $|R_d(x)| \ll 1$ the following result is known:

If $\kappa = O(1)$ as $x \to \infty$,

$$\text{then} \quad S(x,y,\mathcal{P}) \sim X \prod_{\substack{p \leq y \\ p \in \mathcal{P}}} \left(1 - \frac{\omega(p)}{p}\right) \quad \text{holds uniformly} . \tag{8.3}$$

Here the error terms depend only on κ (and x) and not on y or $\mathcal{P}$. For a proof of (8.3) see Iwaniec [12] where there is also a discussion of dimension well removed from zero.

If we interpret the sum (4.9) in terms of the sieve, the analogue of dimension is κ_g that satisfies

$$\prod_{\substack{2 \leq y_1 \leq p \leq y_2 \\ p \in \mathcal{P}}} \left(1 - \frac{\omega(p) g^*(p)}{p}\right)^{-1} \ll \left(\frac{\log y_2}{\log y_1}\right)^{\kappa_g} . \tag{8.4}$$

Since $0 \leq g^* \leq 1$ it is clear from (8.2) and (8.3) that we can always choose

$$\kappa_g \leq \kappa .$$

So in conformity with the general principle stated above we expect to get information about sums

$$S(x,y,\mathcal{P},g) = \sum_{n \leq x, n \in S} a_n \sum_{\substack{d \mid n \\ p \mid d, \Rightarrow p \leq y \text{ and } p \in \mathcal{P}}} \mu(d) g^*(d)$$

from results concerning $S(x,y,\mathcal{P})$, when $0 < g \leq 1$. Viewed from this angle it should not be surprising that (4.10) of Lemma 4 followed from the methods underlying (4.6).

There could however be a loss of information in making the transition from (8.2) to (8.4). For instance consider a situation when $\kappa \gg 1$ but $\kappa_g = o(1)$ by taking f to satisfy

$$\{\max_{p \leq x} f(p)\}/\sqrt{B_2(x)} \to 0, \quad \text{as} \quad x \to \infty \tag{8.5}$$

and setting $\mathcal{P} = \mathcal{P}^{(\infty)}$. When (8.5) holds we know that for f,S satisfying the hypothesis of Theorem 1, the limiting distribution is Gaussian. When (8.5) holds, the function g in (3.4) is ~ 1 and so g^* and hence κ_g are both o(1). So analogous to (8.3) we expect that (4.10) in Lemma 4 will hold for all $y \leqslant x$ in this case and hence consideration of the truncated function $f_y(n)$ should be unnecessary. On the other hand (4.6) would be false at $y = x$.

The above argument should work if instead of (8.5) we had a similar statement of an average sort, with $\kappa_g = o(1)$, characterizing the limiting distribution as Gaussian. To get limiting distributions in Theorem 1 to be non-Gaussian we will have to make the left side of (8.5) $\gg 1$ and so the consideration of f_y is necessary. This interpretation of κ_g in terms of sieve dimension may be worth pursuing further.

§9. LIMITATIONS AND OTHER APPLICATIONS

We have so far restricted the statements of Theorems 1, 2 and 3 to $f \geqslant 0$. This limitation is imposed by the sieve and the classification into Cases 1 and 2. We can treat real valued f using (2.2) but then asymptotic estimates for moments will be replaced by upper bounds of the same order of magnitude. We may then recover the asymptotic estimates from such bounds by appeal to Lemma 2 and a method due to Kubilius which supplies the limiting distribution. It is desirable to remove this dependence on the Kubilius method but the chances at present seem remote since the techniques described in §§4 and 5 apply only to g(n) having their range in $[0,1]$ or $[1,\infty)$. We need a similar treatment of g(n) having $[0,\infty)$ as the range and this looks difficult.

Being based upon the sieve our method can be employed to

estimate moments and obtain distribution theorems for $f(n)$, when $n \in S_{x,y} = \{n \leqslant x \mid p \mid n \Rightarrow p > y\}$, where α is large. Such results which are uniform in y and α are interesting and useful. We shall discuss this on a later occasion.

Finally there is the problem of sums involving $g(n)$, over $n \in S$, where $g(n)$ is permitted to take the value zero also. Throughout this paper, $g(p)$ was always bounded away from zero. When $g(n)$ is summed over $n \in S_{x,y}$, this is indeed a sum over $n \in S$, with $g(p) = 0$ for $p \leqslant y$. Such a sum can be handled by the combinatorial sieve. But if g vanishes over an arbitrary set of primes, then not much is known. It may be worthwhile to see if our method could be adapted to deal with such situations.

ACKNOWLEDGEMENT: I enjoyed the hospitality and financial support of the University of Texas, Austin, while participating in this conference. Also, during the conference I had many fruitful discussions on several aspects of this work with Professors Erdös, Halberstam, Iwaniec, Montgomery and Vaaler.

Department of Mathematics
The University of Texas
Austin, Texas 78712

Permanent Address
MATSCIENCE
The Institute of Mathematical Sciences
Madras 600113, INDIA

REFERENCES

[1] Alladi, K., A study of the moments of additive functions using Laplace transforms and sieve methods, Proc. Fourth Matscience Conf. On Number Th., Ootacamund, India (1984), Springer Lecture Notes (to appear).

[2] ------, Moments of additive functions and sieve methods, Proc. of the New York Number Theory Seminar, 1982, Springer Lecture Notes, Vol. 1051, 1 - 25.

[3] Delange, H., Sur un théorème d'Erdös et Kac, Acad. Roy. Belg. Bull. Cl. Sci., 5, 42(1956), 130 - 144.

[4] ------, Sur les fonctions arithmetiques fortement additives, C.R. Acad. Sci. Paris, 244(1957),2122-2124 .

[5] Elliott,P.D.T.A., Probabilistic Number Theory, Vol. 1 and 2, Grundelehren 239 - 240, Springer-Verlag, Berlin, New York (1980).

[6] ------, High power analogues of the Turán-Kubilius inequality and an application to number theory, Can. J. Math., 32(1980), 893 - 907.

[7] Erdös, P. and Kac, M., The Gaussian law of errors in the theory of additive number theoretic functions, Amer. J. Math., 61(1939), 713 - 721.

[8] Halberstam, H., On the distribution of additive number theoretic functions I, II, III, J. London Math. Soc., 30(1956), 43 - 53; 31(1956), 1 - 14; 31(1956), 14 - 27.

[9] Halberstam, H. and Richért, H.-E., Sieve Methods, Academic Press, London-New York (1974).

[10] Hall, R.R., Halving an estimate obtained from Selberg's upper bound method, Acta Arith., 25(1974), 347 - 351.

[11] Hall, R.R. and Tenenbaum, G., On the local behavior of some arithmetical functions, Acta Arith., 43(1984), to appear.

[12] Iwaniec, H., Rosser's Sieve, Acta Arith., 36(1980), 171 - 202.

[13] Kubilius, J., Probabilistic methods in the theory of numbers, Amer. Math. Soc., Translations of Mathematical Monographs 11, Providence (1964).

On the Distribution of the Zeros of the Riemann Zeta-Function

J. B. Conrey

In this article we shall describe recent results concerning the zeros of the zeta-function. In particular we are interested in the proportion of zeros of $\zeta(s)$ on the ½-line, the proportion of simple zeros of $\zeta(s)$ on the ½-line, and extreme gaps between the ordinates of consecutive zeros of $\zeta(s)$. Some of the results we quote are conditional; this will be appropriately indicated.

Selberg [19] was the first to show that a positive proportion of the zeros of $\zeta(s)$ are on $\sigma = \frac{1}{2}$. Levinson [11], by a different method, showed that this proportion, which we shall denote α, exceeds 0.3420. We briefly describe Levinson's method and then indicate modifications of it which have led to small improvements.

The starting point of Levinson's method is the identity

$$H(s)\zeta(s) = H(s)G(s) + H(1-s)G(1-s) \tag{1}$$

where

$$H(s) = \tfrac{1}{2}\, s(s-1)\pi^{-s/2}\,\Gamma\left(\frac{s}{2}\right)$$

and

$$G(s) = \zeta(s) - \frac{\zeta'(s)}{\frac{\chi'}{\chi}(s)} . \tag{2}$$

Here

$$\frac{\chi'}{\chi}(s) = -\log\frac{|t|}{2\pi} + O\left(\frac{1}{|t|}\right) \tag{3}$$

uniformly for $|\sigma| \leqslant 10$, $|t| \geqslant 1$. It follows that $\zeta(\frac{1}{2} + it) = 0$ when

$$\arg HG(\tfrac{1}{2} + it) \equiv \frac{\pi}{2} \bmod \pi \ . \tag{4}$$

Now

$$\arg H(\tfrac{1}{2} + it) = \frac{t}{2} \log \frac{|t|}{2\pi e} + O(1) \tag{5}$$

so that a bound of the shape

$$\left| \arg G(\tfrac{1}{2} + it) \Big|_{t=T}^{T+U} \right| \leqslant \beta \frac{UL}{2} \tag{6}$$

where

$$L = \log \frac{T}{2\pi} \ , \quad U = TL^{-10} \tag{7}$$

implies that

$$\alpha \geqslant 1 - \beta \ . \tag{8}$$

(It is easier to work on the interval $[T, T+U]$ than on $[0, T]$.) To obtain the bound (6) we use the argument principal on the rectangle with vertices $2 + iT$, $2 + i(T+U)$, $\tfrac{1}{2} + i(T+U)$, $\tfrac{1}{2} + iT$. This leads to (6) with

$$\beta UL = 4\pi N_G \tag{9}$$

where N_G is the number of zeros of $G(s)$ in $\sigma \geqslant \tfrac{1}{2}$, $T \leqslant t \leqslant T + U$. To bound N_G we apply Littlewood's lemma to ψG on the rectangle with vertices $2 + iT$, $2 + i(T+U)$, $a + iT$ where

$$\psi(s) = \sum_{n \leqslant y} b(n) n^{-s} \tag{10}$$

is a Dirichlet polynomial with $b(1) = 1$, $b(n) \ll 1$, and

$$a = \frac{1}{2} - \frac{R}{L} \tag{11}$$

for some fixed $R > 0$. This leads to

$$2\pi\,(\tfrac{1}{2} - a)N_G \leqslant \int_T^{T+U} \log|\psi G(a+it)|\,dt$$
$$\leqslant \frac{U}{2}\log\Big(\frac{1}{U}\int_T^{T+U} |\psi G(a+it)|^2 dt\Big) \tag{12}$$

so that

$$\alpha \geqslant 1 - \frac{\log\big(\frac{1}{U}\int_T^{T+U} |\psi G(a+it)|^2 dt\big)}{R} . \tag{13}$$

Levinson evaluated the integral in (13) in the case that

$$y = T^{\frac{1}{2}} L^{-20} \tag{14}$$

and

$$b(n) = \frac{\mu(n)}{n^{\frac{1}{2}-a}}\ \frac{\log y/n}{\log y} \tag{15}$$

and obtained

$$\frac{1}{U}\int_T^{T+U} |\psi G(a+it)|^2 dt \sim F(R) \tag{16}$$

where

$$F(R) = e^{2R}\Big(\frac{1}{2R^3} + \frac{1}{24R}\Big) - \frac{1}{2R^3} - \frac{1}{R^2} - \frac{25}{24R} + \frac{7}{12} - \frac{R}{12} . \tag{17}$$

With $R = 1.3$ this led to

$$\alpha \geqslant 0.3420 . \tag{18}$$

Subsequently, in an attempt to optimize the coefficients of the mollifier $\psi(s)$, Levinson [13] was led to the choice

$$b(n) = \frac{\mu(n)}{n^{1-2a}}\ \frac{y^{1-2a} - j^{1-2a}}{y^{1-2a} - 1} . \tag{19}$$

This choice gives the result

$$\alpha \geqslant 0.3474 . \tag{20}$$

Heath-Brown [9] and Selberg independently noticed that the zeros located by Levinson's method are simple zeros of $\zeta(s)$. Thus

$$\alpha_s \geqslant 0.3474 \tag{21}$$

where α_s is the proportion of simple zeros of $\zeta(s)$ on the ½-line.

Lou [14] chose the mollifier $\Psi(s) = \psi(s) + \chi(s)L^2 h\, \psi_1(s)$ where ψ is as in (10) and (19), h is constant, and

$$\psi_1(s) = \sum_{k \leqslant y} \frac{\mu^2(k) b_1(k)}{k^{\frac{1}{2}+a-s}} \frac{\log y/k}{\log y}, \quad b_1(k) = \sum_{n|k} \frac{\mu(n) d(n)}{n^{2s-2a}}. \tag{22}$$

This leads to

$$\alpha, \alpha_s \geqslant 0.35. \tag{23}$$

Levinson [12] suggested new identities for $\zeta(s)$ which generalize (1) and lead to better results. In $[T, T+U]$ we can write Levinson's G as

$$G(s) \sim \zeta(s) + \frac{\zeta'(s)}{L} \tag{24}$$

because of (3). Conrey [4] used identities of the sort

$$H(s)\zeta(s) = H(s)\mathfrak{G}(s) + H(1-s)\mathfrak{G}(1-s) \tag{25}$$

where

$$\mathfrak{G}(s) \sim \sum_n a_n \zeta^{(n)}(s) L^{-n} \tag{26}$$

and the a_n are certain complex numbers. These identities can be derived in a variety of ways. (See Levinson [11] and [12], Bombieri [3], Conrey [4] and [5], and Anderson [1] for different approaches.) Still another method for developing (25) and (26) in a rather simple way is as follows.

Let g_0 be arbitrary, g_{2r} purely imaginary, and g_{2r+1} real for $r = 1, 2, \ldots$. Let

$$\xi(s) = H(s)\zeta(s) \tag{27}$$

so that $\xi(\frac{1}{2} + it)$ is real and $\xi(s) = \xi(1-s)$ is the functional equation for $\zeta(s)$. Then

$$\begin{aligned}(2 \operatorname{Re} g_0)\xi(s) &= g_0\xi(s) + \overline{g}_0\xi(1-s) \\ &= g_0\xi(s) + \sum_r g_r\xi^{(r)}(s)L^{-r} \\ &\quad + \overline{g}_0\xi(1-s) - \sum_r g_r(-1)^r\xi^{(r)}(1-s)L^{-r}\end{aligned} \tag{28}$$

since

$$\xi^{(r)}(s) = (-1)^r\xi^{(r)}(1-s) . \tag{29}$$

This expresses $(2 \operatorname{Re} g_0)\xi(s)$ as a sum of complex conjugates when $s = \frac{1}{2} + it$, just as in (1). We take

$$\mathfrak{G}(s) = \Big(g_0\xi(s) + \sum_r g_r\xi^{(r)}(s)L^{-r}\Big)/H(s) . \tag{30}$$

We rewrite $\xi^{(r)}(s)$ using (27) and Leibniz's formula. Also

$$\frac{H^{(k)}(s)}{H(s)} \sim \left(\frac{L}{2}\right)^k \tag{31}$$

for $T \leqslant t \leqslant T + U$. Thus

$$\mathfrak{G}(s) \sim g_0\zeta(s) + \sum_r g_rL^{-r} \sum_k \binom{r}{k} \zeta^{(k)}(s)\left(\frac{L}{2}\right)^{r-k} \tag{32}$$

which is (26) with

$$a_0 = g_0 + \sum_r 2^{-r}g_r \tag{33}$$

and

$$a_k = 2^k \sum_r 2^{-r} \binom{r}{k} g_r \tag{34}$$

for $k \geqslant 1$. Hence the a_k are restricted only by the conditions on the g_r. Since g_0 is arbitrary, we may take $a_0 = 1$. With this normalization, (13) holds with $\mathfrak{G}$ in place of G.

Conrey [4] also used the more general mollifier coefficients

$$b(n) = \frac{\mu(n)}{n^{\frac{1}{2} - a}} P\left(\frac{\log {}^{y}/n}{\log y} \right) \tag{35}$$

where P is an analytic function with $P(0) = 0$ and $P(1) = 1$. By choosing the a_k and P appropriately this gives [4]

$$\alpha \geqslant 0.3658 . \tag{36}$$

(It is possible to choose P optimally here by calculus of variations.)

With Levinson's original G and the mollifier coefficients (35) one can show [5]

$$\alpha_s \geqslant 0.3485 \tag{37}$$

which is not as good as (23). However, Anderson [1] pointed out that for counting multiple zeros one may use

$$\mathfrak{G}_1(s) = \zeta(s) + a_1 \zeta'(s) L^{-1} \tag{38}$$

with a_1 an arbitrary real. He used this and the coefficients (19) and obtained [1]

$$\alpha_s \geqslant 0.3532 . \tag{39}$$

If one uses (38) with the coefficients (35) the result is

$$\alpha_s \geq 0.358 \ . \tag{40}$$

Further improvements in the method seem to rely on taking a longer Dirichlet polynomial for the mollifier. That is, we want

$$y = T^{\theta}, \quad \theta > 1/2 \ . \tag{41}$$

Balasubramanian, Conrey, and Heath-Brown [2] have shown (using (10) and (35) with $a = 1/2$) that

$$\frac{1}{T}\int_0^T |\zeta(\tfrac{1}{2} + it)|^2 \, |\psi(\tfrac{1}{2} + it)|^2 dt \sim 1 + \frac{\log T}{\log y}\int_0^1 P'(x)^2 dx \tag{42}$$

for any $\theta < 9/17$. This result needs to be generalized to the integrals

$$\frac{1}{T}\int_0^T \zeta^{(u)}(a + it)\zeta^{(v)}(a - it) \, |\psi(\tfrac{1}{2} + it)|^2 dt$$

in order to evaluate (16). If this is done, one will obtain the result

$$\alpha \geq 0.38 \ . \tag{43}$$

If Hooley's conjecture R^* [10] is assumed, then one can prove (42) for any $\theta < 4/7$ which would lead to

$$\alpha \geq 0.4077 \ . \tag{44}$$

Next we consider what bounds we can obtain for α_s if we first assume something about the zeros of $\zeta(s)$. Montgomery [15] showed, assuming the Riemann Hypothesis, that

$$\text{(R/H)} \qquad \alpha_s \geq 2/3 \ . \tag{45}$$

With Taylor (see [16]), he improved this to

$$\text{(RH)} \qquad \alpha_s \geqslant \frac{3}{2} - \frac{\sqrt{2}}{2} \cot \frac{\sqrt{2}}{2} = 0.6725 . \tag{46}$$

Assuming his pair correlation conjecture, Montgomery [15] obtained

$$\text{(RH, PC)} \qquad \alpha_s = 1 . \tag{47}$$

Recently, by a new technique, Conrey, Ghosh, and Gonek [6] showed, assuming the generalized Riemann Hypothesis, that

$$\text{(GRH)} \qquad \alpha_s \geqslant \frac{19}{27} = 0.703\ldots . \tag{48}$$

The assumption can probably be weakened to RH. The method we used is as follows.

Assume the Riemann Hypothesis. Let $\frac{1}{2} + i\gamma$ denote a typical zero of $\zeta(s)$. Then by Cauchy's inequality

$$N_s(T) \geqslant \frac{\left| \sum_{\gamma \leqslant T} \psi\zeta'(\frac{1}{2} + i\gamma) \right|^2}{\sum_{\gamma \leqslant T} \left| \psi\zeta'(\frac{1}{2} + i\gamma) \right|^2} \tag{49}$$

where $N_s(T)$ is the number of simple zeros of $\zeta(s)$ with $0 < \gamma \leqslant T$ and $\psi(s)$ is as in (10), (14), and (35) with $a = \frac{1}{2}$. The expressions on the right side of (49) can be estimated asymptotically by methods similar to, but more complicated than what Gonek [8] used. For a real function $P(x)$ the numerator is

$$\sim \left(\frac{1}{2} + \frac{1}{2} \int_0^1 P(x)\,dx \right)^2 \left(\frac{TL^2}{2\pi} \right)^2 \tag{50}$$

and the denominator is

$$\sim \left(\frac{1}{3} + \frac{1}{4} \left(\int_0^1 P(x)\,dx \right)^2 + \frac{1}{2} \int_0^1 P(x)\,dx + \frac{1}{6} \int_0^1 P'(x)^2 dx \right) \frac{TL^3}{2\pi} . \tag{51}$$

By the calculus of variations

$$P(x) = -\frac{1}{2}x^2 + \frac{3}{2}x \qquad (52)$$

is optimal from which we obtain (48). We needed GRH to obtain the estimate (51) but his could probably be done by another method.

It is interesting that the integral on the left side of (42) arises in the evaluation (51). It may be possible, as with (42), to take $y = T^\theta$ with $\theta > 1/2$ in this problem which would lead to an improvement in (48).

The last problem we consider is the existence of small and large gaps between the ordinates of consecutive zeros of $\zeta(s)$. Let γ and γ' denote ordinates of consecutive zeros of $\zeta(s)$ with $0 < \gamma \leqslant \gamma'$. Then the average value of

$$(\gamma' - \gamma)\frac{\log \gamma}{2\pi} \qquad (53)$$

is 1. Let

$$\lambda = \limsup\,(\gamma' - \gamma)\frac{\log \gamma}{2\pi} \qquad (54)$$

and

$$\mu = \liminf\,(\gamma' - \gamma)\frac{\log \gamma}{2\pi}\,. \qquad (55)$$

Selberg [20] remarks that $\mu < 1$ and $\lambda > 1$ can be shown (unconditionally). Montgomery [15] obtained

$$(\mathrm{RH}) \qquad \mu < 0.68 \qquad (56)$$

while Mueller [18], using results from Gonek's thesis (see [8]) obtained

$$(\mathrm{RH}) \qquad \lambda > 1.9\,. \qquad (57)$$

Montgomery and Odlyzko [17] showed

$$\text{(RH)} \qquad \mu < 0.5179\,, \quad \lambda > 1.9799\,. \tag{58}$$

Recently, Conrey, Ghosh, and Gonek [7] have proven

$$\text{(RH)} \qquad \mu < 0.5172\,, \quad \lambda > 2.337\,. \tag{59}$$

Our idea is based on that of Mueller [18] and has surprising similarity to the method of Montgomery and Odlyzko [17] which appears to be much different at the outset. We assume the Riemann Hypothesis and consider

$$M_1 = \int_{-\beta}^{\beta} \sum_{T < \gamma \leqslant 2T} |A(\tfrac{1}{2} + i \quad + i\alpha)|^2 d\alpha \tag{60}$$

where $\beta = \pi b/L$, $L = \log T$, and

$$A(s) = \sum_{n \leqslant N} a(n) n^{-s} \tag{61}$$

is a Dirichlet polynomial of length

$$N = T^{1-\delta} \tag{62}$$

where $\delta > 0$ is small. We compare this to

$$M_2 = \int_{T}^{2T} |A(\tfrac{1}{2} + it)|^2 dt\,. \tag{63}$$

If for some choice of A,

$$M_1 < M_2 \tag{64}$$

then $\zeta(s)$ has a pair of consecutive zeros with ordinates in $[T, 2T]$ which are farther apart than $2\pi b/L$, that is, farther apart than b times the average. If

$$M_2 < M_1 \tag{65}$$

then $\zeta(s)$ has a pair of consecutive zeros with ordinates in $[T, 2T]$ which are nearer to each other than b times the average. We can carry out the estimation of M_1 and M_2 asymptotically for any arithmetical function $a(n)$ with

$$a(n) \ll_{\varepsilon} n^{\varepsilon} . \tag{66}$$

This leads to a formula which is equivalent to (20) and (21) of Montgomery and Odlyzko [17]. To obtain long gaps we take

$$a(n) = d_r(n) , \tag{67}$$

the coefficient of n^{-s} in the Dirichlet series for

$$\zeta(s)^r \tag{68}$$

with $r = 2.2$. To obtain short gaps we take

$$a(n) = \lambda(n) d_r(n) \tag{69}$$

where λ is Liouville's function and $r = 1.1$. This leads to (59).

School of Mathematics
The Institute for Advanced Study
Princeton, New Jersey 08540

Present address:

Department of Mathematics
Oklahoma State University
Stillwater, Oklahoma 74074

REFERENCES

1. Anderson, R.J., Simple zeros of the Riemann zeta-function, J. Number Th., 17 (1983), 176-182.
2. Balasubramanian, R., J.B. Conrey, and D.R. Heath-Brown, Asymptotic mean square of the product of the Riemann zeta-function and a Dirichlet polynomial, J. Reine Angew. Math., to appear.
3. Bombieri, E., A lower bound for the zeros of Riemann's zeta-function on the critical line (following N. Levinson), Seminaire Bourbaki 469 (1975), 176-181.
4. Conrey, J.B., Zeros of derivatives of Riemann's xi-function on the critical line, J. Number Th., 16 (1983), 49-74.
5. ------, Zeros of derivatives of Riemann's xi-function on the critical line II, J. Number Th., 17 (1983), 71-75.
6. ------, A. Ghosh, and S.M. Gonek, Simple zeros of the zeta function, preprint.
7. ------, ------, and ------, A note on gaps between zeros of the zeta function, Bull. London Math. Soc., 16 (1984), 421-424.
8. Gonek, S.M., Mean values of the Riemann zeta-function and its derivatives, Invent. Math., 75 (1984), 123-141.
9. Heath-Brown, D.R., Simple zeros of the Riemann zeta-function on the critical line, Bull. London Math. Soc. 11 (1979), 17-18.
10. Hooley, C., On the greatest prime factor of a cubic polynomial, J. Reine Angew. Math., 303/304 (1978), 21-50.
11. Levinson, N., More than one-third of zeros of Riemann's zeta-function are on $\sigma = 1/2$, Adv. Math. 13 (1974), 383-436.
12. ------, Generalization of recent method giving lower bound for $N_0(T)$ of Riemann's zeta-function, Proc. Nat. Acad. Sci. USA 71, No. 10 (1974), 3984-3987.

13. ------, Deduction of semi-optimal mollifier for obtaining lower bound for $N_0(T)$ for Riemann's zeta-function, Proc. Nat. Acad. Sci. USA 72, No. 1 (1975), 294-297.

14. Lou Shi-Tuo, A lower bound for the number of zeros of Riemann's zeta-function on $\sigma = 1/2$, Recent Progress in Analytic Number Theory, Academic Press, London, 1981; vol. 1, 319-324.

15. Montgomery, H.L., The pair correlation of zeros of the zeta function, Proc. Sympos. Pure Math. 24 (1973), 181-193.

16. ------, Distribution of the zeros of the Riemann zeta function, Proceedings of the International Congress for Mathematicians, Vancouver, B.C., 1974, vol. 1, 379-381.

17. ------ and A. Odlyzko, Gaps between zeros of the zeta function, to appear.

18. Mueller, J.H., On the difference between the consecutive zeros of the Riemann zeta function, J. Number Theory, 14 (1982), 327-331.

19. Selberg, A., On the zeros of Riemann's zeta-function, Skr. Norske, Vid. Acad. Oslo c No. 10 (1942), 1-59.

20. ------, The zeta-function and the Riemann hypothesis, Skandinaviske Mathematiker Knogres 10 (1946), 187-200.

Notes added in proof

We can now give a simpler proof of (16) which avoids the use of the approximate functional equation for $G(s)$. Instead, it relies on the determination of the poles of a function similar to Estermann's $\Sigma d(n)e(nh/k)n^{-s}$ for the main terms and the large sieve for the error terms (see [21]).

An even easier approach to [30] is as follows. Suppose that $\eta(s)$ is entire and is imaginary on the $\frac{1}{2}$-line and that g_0 is a complex number which is not purely imaginary. Then, on the $\frac{1}{2}$- line, $g_0 \, \xi(s) + \eta(s)$ is imaginary if and only if $\xi(s) = 0$, since $\xi(\frac{1}{2} + it)$ is real. We take $\eta(s) = \Sigma g_n \xi^{(n)}(s) L^{-n}$ where g_n is real if n is odd and g_n is imaginary if n is even. Then it suffices to bound the change in argument of $(g_0 \, \xi(s) + \eta(s))/H(s)$ as in (30).

We have succeeded in proving (48) subject only to RH. The proof of this is similar to Vaughan's proof of Bombieri's theorem in that it uses an analogous identity and large sieve estimates; however, here we must bound mean sixth powers of L-functions at one stage. By a similar method we can show that at least 1/3 of the zeros of the Riemann zeta-function are not zeros of any given Dirichlet L-function (on RH). In conjunction with (48) this implies that a positive proportion (at least 1/54) of the zeros of the Dedekind zeta-function of a quadratic field are simple (on RH). (See [24].)

Combining the techniques of [6] and [7] we can now improve (59) to $\lambda > 2.68$ on RH. To do this we replace A(s) of (61) by $\zeta(s) \Sigma n^{-s}$ where the sum is for $n \leqslant T^{\theta}$ and $\theta < \frac{1}{2}$. (See [23].)

In [22] we give upper and lower bounds for the proportion of gaps between consecutive zeros of the zeta-function which are less than α times the average spacing. These bounds are non-trivial for $0.77 < \alpha < 1.33$.

21. Conrey, J.B., and A. Ghosh, A simpler proof of Levinson's theorem, Proc. Cambridge Phil. Soc., to appear.

22. ------, ------, D. Goldston, S.M. Gonek, and D.R. Heath-Brown, Distribution of gaps between zeros of the zeta-function, Quarterly J. of Math., to appear.

23. ------, ------, and S.M. Gonek, Large gaps between zeros of the zeta-function, preprint.

24. ------, ------, and ------, Simple zeros of the Dedekind zeta-function of a quadratic field, in preparation.

Remarks on the Difference of Consecutive Numbers Prime to q

Harold G. Diamond and Jeffrey D. Vaaler

1. INTRODUCTION

Let $q \geqslant 2$ be an integer and let

$$(1.1) \qquad 1 = a_1 < a_2 < \ldots < a_{\varphi(q)} = q - 1$$

be the $\varphi(q)$ integers in the interval $[1,q]$ which are prime to q. It will be convenient to set $a_{\varphi(q)+1} = q+1$. Since

$$(1.2) \qquad \sum_{j=1}^{\varphi(q)} (a_{j+1} - a_j) = q$$

we have

$$q^2 \varphi(q)^{-1} \leqslant \sum_{j=1}^{\varphi(q)} (a_{j+1} - a_j)^2$$

by Cauchy's inequality. In 1940 P. Erdös [2] conjectured that

$$(1.3) \qquad \sum_{j=1}^{\varphi(q)} (a_{j+1} - a_j)^2 \ll q^2 \varphi(q)^{-1}$$

and this remains a surprisingly difficult open problem.[2]

The first results in the direction of (1.3) were obtained by Hooley [5]. He showed that if $1 \leqslant \alpha < 2$ then

$$(1.4) \qquad \sum_{j=1}^{\varphi(q)} (a_{j+1} - a_j)^{\alpha} \ll_{\alpha} q\left(\frac{q}{\varphi(q)}\right)^{\alpha-1} ,$$

where our notation $\ll_{\alpha}$ indicates that the implied constant depends on α. By applying Hölder's inequality to (1.2) one easily sees that (1.4) is best possible aside from a determination of the constant. For $\alpha = 2$ Hooley was able to establish the weaker bound

$$(1.5) \qquad \sum_{j=1}^{\varphi(q)} (a_{j+1} - a_j)^2 \ll \left(\frac{q^2}{\varphi(q)}\right) (\log \log q) .$$

By making a more careful computation in Hooley's method, M. Hausman and H. N. Shapiro [4] obtained the slightly sharper inequality

$$(1.6) \qquad \sum_{j=1}^{\varphi(q)} (a_{j+1} - a_j)^2 \ll \left(\frac{q^2}{\varphi(q)}\right) (1 + \sum_{p|q} p^{-1} \log p) ,$$

where the sum on the right of (1.6) is over prime divisors of q. It follows from (1.6) that for any $\epsilon > 0$ there exists a constant $M(\epsilon) > 0$ such that

$$\sum_{j=1}^{\varphi(q)} (a_{j+1} - a_j)^2 \leq M(\epsilon)\, q^2 \varphi(q)^{-1}$$

fails to hold on a set of integers q which has an upper density less than ϵ. The bound (1.6) was also discovered independently by R. C. Vaughan and by K. Norton [7].

For $1 \leq \alpha < 2$, Hooley [6] has established an asymptotic distribution for the $\varphi(q)$ terms $(a_{j+1} - a_j)^\alpha$. This gives more precise information than (1.4) but does not seem to shed new light upon the conjecture (1.3). If $2 < \alpha$ there are evidently no nontrivial bounds known for the sum on the left of (1.4). We note, however, that Erdös [3] has conjectured that (1.4) holds for all $\alpha \geq 1$.

Our purpose here is to give a more general form of Hooley's method which leads to an extremal problem for a certain Hermitian form. The solution to this extremal problem, as well as a simpler estimate of the solution, leads to an improvement in the bound (1.6). Unfortunately, the improved bound which we obtain is quite complicated and we are unable to decide if it implies the conjectured inequality (1.3). Thus, our results should be viewed as the introduction of a new technique which is not yet fully understood. Since it seems to be of some independent interest, we discuss a slightly more general Hermitian form than we need for improving (1.6). Our investigation of this form makes use of some recent results for the finite Fourier transform which we obtained in [1].

We briefly summarize our work. Theorem 1, which is almost trivial, shows that it suffices to consider the sums

$$\sum_{j=1}^{\varphi(q)} (a_{j+1} - a_j)^{\alpha}$$

for odd and square free values of q. This is a useful simplification which allows us to appeal directly to the results in [1]. Theorem 2 is our modification of Hooley's method in which the upper bound depends on the values of certain Hermitian forms. We have restricted our attention to the case $\alpha = 2$. In Theorem 4, we give an upper bound for the minimum value of the Hermitian forms over an affine subspace. This uses in a critical way certain results obtained in [1]. Theorem 2 and Theorem 4 are then combined to produce an upper bound which is at least as sharp as (1.6). Specifically, the arithmetic mean of a nonconstant function has been replaced by the harmonic mean.

2. A REFINEMENT OF HOOLEY'S METHOD

Let $S_\alpha(q)$ denote the left hand side of (1.4) and then set

$$T_\alpha(q) = q^{-1} \left(\frac{q}{\varphi(q)}\right)^{1-\alpha} S_\alpha(q) \quad . \tag{2.1}$$

THEOREM 1. If d is the largest odd square free divisor of q then $T_\alpha(d) = T_\alpha(q)$.

Proof. Let a_j be the integers satisfying (1.1) and let u be the largest square free divisor of q. Clearly the integers

$$1 = a_1 < a_2 < \ldots < a_{\varphi(u)} = u - 1$$

(and $a_{\varphi(u)+1} = u+1$) are exactly the integers in $[1,u]$ which are prime to u. Since an integer is prime to u if and only if it is prime to q, we also find that

$$a_{k\varphi(u)+j} = ku + a_j$$

for $k = 0, 1, 2, \ldots, (q/u) - 1$ and $j = 1, 2, \ldots, \varphi(u)$. It follows that

$$(2.2) \qquad S_\alpha(q) = \sum_{k=0}^{(q/u)-1} \sum_{j=1}^{\varphi(u)} (a_{k\varphi(u)+j+1} - a_{k\varphi(u)+j})^\alpha$$

$$= (q/u)\, S_\alpha(u) \quad .$$

We now combine (2.1), (2.2) and the identity $\frac{u}{\varphi(u)} = \frac{q}{\varphi(q)}$ to conclude that $T_\alpha(u) = T_\alpha(q)$.

If q is odd we are done. But if q is even then $2d = u$ and an additional argument is necessary. Let

$$1 = b_1 < b_2 < \ldots < b_{\varphi(d)} = d - 1$$

be the integers in $[1,d]$ which are prime to d and set $b_{\varphi(d)+1} = d+1$. Obviously, the integers $2b_j - d$, $j = 1, 2, \ldots, \varphi(d)$, are prime to $2d$.

Since

$$(2b_{\varphi(d)+1} - d) - (2b_1 - d) = 2d$$

and $\varphi(d) = \varphi(2d)$, the set $\{2b_j - d,\ j = 1, 2, \ldots, \varphi(d)\}$ must consist of $\varphi(2d)$ consecutive integers prime to $2d$. Therefore we have

$$(2.3) \qquad S_\alpha(2d) = \sum_{j=1}^{\varphi(d)} \{(2b_{j+1} - d) - (2b_j - d)\}^\alpha = 2^\alpha S_\alpha(d) \quad .$$

The identity $T_\alpha(d) = T_\alpha(2d) = T_\alpha(u)$ now follows from (2.1) and (2.3).

Next we introduce some notation. Let $\mathbb{Z}_q$ denote the ring of residue classes of integers modulo q and let $G(q) \subseteq \mathbb{Z}_q$ be the multiplicative group of residue classes which are relatively prime to q. If $f: \mathbb{Z}_q \to \mathbb{C}$ we define the (finite) Fourier transform of f to be the function $f: \mathbb{Z}_q \to \mathbb{C}$ determined by

$$(2.4) \qquad \hat{f}(n) = q^{-1/2} \sum_{m=1}^{q} f(m) e\left(\frac{-nm}{q}\right) ,$$

where $e(x) = e^{2\pi i x}$. If f and g are both functions from $\mathbb{Z}_q$ into C we define their convolution $f*g: \mathbb{Z}_q \to C$ by

$$(2.5) \qquad f*g(m) = q^{-1/2} \sum_{n=1}^{q} f(m-n)g(n) \quad .$$

These operations are related by the well-known identity

$$(2.6) \qquad \widehat{f*g}(n) = \hat{f}(n)\hat{g}(n) \quad .$$

We will also require the Parseval formula

(2.7) $$\sum_{n=1}^{q} |\hat{f}(n)|^2 = \sum_{m=1}^{q} |f(m)|^2 .$$

Functions $f: \mathbb{Z}_q \to \mathbb{C}$ having support contained in the group $G(q)$ will be of particular importance. We define $\mathfrak{G}_q$ to be the class of functions $f: \mathbb{Z}_q \to \mathbb{C}$ satisfying

(2.8) the support of f is contained in $G(q)$, that is, $f(m) = 0$ if $(m,q) > 1$,

and

(2.9) $$\sum_{m=1}^{q} f(m) = \varphi(q) .$$

Now let h be an integer, $1 \leqslant h \leqslant q-1$. If $f \in \mathfrak{G}_q$ we define

(2.10) $$Q(f,h) = \sum_{m=1}^{q} \left| \sum_{n=m+1}^{m+h} f(n) - \frac{h\varphi(q)}{q} \right|^2 .$$

It will also be convenient to express $Q(f,h)$ using the convolution (2.5). We do this by using the functions

(2.11) $$K_h(m) = \begin{cases} 1 - (h/q) & \text{if } m \equiv -1, -2, \dots, -h \bmod q , \\ -(h/q) & \text{otherwise} . \end{cases}$$

It follows easily that

(2.12) $$Q(f,h) = q \sum_{m=1}^{q} \left| f*K_h(m) \right|^2$$

whenever $f \in \mathfrak{G}_q$.

An example of a function f in the class $\mathfrak{G}_q$ is the principal Dirichlet character mod q, which we denote by $\chi = \chi_q$. In [5] Hooley obtained the bound (1.5) by estimating, in effect, $Q(\chi,h)$ and then using this estimate to bound $S_2(q)$. The following inequality provides a slightly more flexible way of proceeding. Here we have used the notation

$$M(q) = \max_{1 \leqslant j \leqslant \varphi(q)} (a_{j+1} - a_j) .$$

THEOREM 2. *For each integer* h, $1 \leqslant h \leqslant M(q)$, *let* f_h *be an arbitrary function in* $\mathfrak{G}_q$. *Then*

$$S_2(q) \leqslant q + 2\left(\frac{q}{\varphi(q)}\right)^2 \sum_{h=1}^{M(q)} h^{-2} Q(f_h,h) . \tag{2.13}$$

Proof. Suppose that for some integer j, $a_{j+1} - a_j \geqslant h + 1$. Then

$$\left| \sum_{n=a_j+s+1}^{a_j+s+h} f_h(n) - \frac{h\varphi(q)}{q} \right|^2 = \left(\frac{h\varphi(q)}{q}\right)^2$$

for $s = 0,1,2,\ldots, a_{j+1} - a_j - h - 1$. Therefore we have

$$Q(f_h,h) = \sum_{m=1}^{q} \left| \sum_{n=m+1}^{m+h} f_h(n) - \frac{h\varphi(q)}{q} \right|^2$$

$$> \left(\frac{h\varphi(q)}{q}\right)^2 \sum_{j=1}^{\varphi(q)} (a_{j+1} - a_j - h)^+ ,$$

where $x^+ = \max\{x,0\}$. It follows that

$$\left(\frac{q}{\varphi(q)}\right)^2 \sum_{h=1}^{M(q)} h^{-2} Q(f_h,h) \tag{2.14}$$

$$\geqslant \sum_{h=1}^{M(q)} \sum_{j=1}^{\varphi(q)} (a_{j+1} - a_j - h)^+$$

$$= \sum_{j=1}^{\varphi(q)} \sum_{h=1}^{a_{j+1}-a_j} (a_{j+1} - a_j - h)$$

$$= \frac{1}{2} \sum_{j=1}^{\varphi(q)} (a_{j+1} - a_j)(a_{j+1} - a_j - 1)$$

$$= \frac{1}{2} S_2(q) - \frac{1}{2} q \ .$$

From (2.14) we obtain the bound (2.13).

3. THE EXTREMAL PROBLEM

It will be convenient to consider the inequality (2.13) in the form

$$(3.1) \qquad T_2(q) \ll 1 + \varphi(q)^{-1} \sum_{h=1}^{M(q)} h^{-2} Q(f_h, h) \ .$$

Our next step is to select functions $f_h \in \mathfrak{G}_q$ so that $Q(f_h, h)$ is small. In view of Theorem 1 we may restrict ourselves to square free q (and if necessary to odd, square free q.) We therefore formulate the following general problem. Let $q \geqslant 2$ be square free and let $K: \mathbb{Z}_q \to \mathbb{C}$. Define a non-negative Hermitian form in the q complex variables $f(1), f(2), \ldots, f(q)$ by

$$(3.2) \qquad Q(f,K) = q \sum_{m=1}^{q} |f{*}K(m)|^2$$

and minimize $Q(f,K)$ over functions f in $\mathfrak{G}_q$ with K

fixed. Applying (2.6) and (2.7), we see that

$$Q(f,K) = q \sum_{n=1}^{q} |\hat{f}(n)|^2 |\hat{K}(n)|^2 , \tag{3.3}$$

which diagonalizes the form. Now, however, it is not immediately clear how the values of $\hat{f}(n)$ in (3.3) should be determined so as to insure that $f \in \mathfrak{C}_q$. To overcome this difficulty we appeal to the results in [1], which we briefly describe.

If d is a positive divisor of the square free integer q we define $\bar{d}$ to be the unique integer, $1 \leqslant \bar{d} \leqslant (q/d)$, which satisfies $d\bar{d} \equiv 1 \bmod (q/d)$. Thus $\{d\bar{d} : d|q\}$ is precisely the set of idempotent elements in the ring $\mathbb{Z}_q$. In particular the residue classes $d\bar{d} \bmod q$ are distinct since $(d\bar{d}, q) = d$. We define the function $W: \mathbb{Z}_q \to \mathbb{C}$ by

$$W(m) = \sum_{d|q} \mu(d)\, e\left(\frac{d\bar{d}m}{q}\right) , \tag{3.4}$$

where μ denotes the Möbius function. From (3.4) we see immediately that

$$\hat{W}(n) = \begin{cases} q^{1/2}\mu(d) & \text{if } n \equiv d\bar{d} \bmod q \text{ for some } d|q , \\ 0 & \text{otherwise.} \end{cases} \tag{3.5}$$

If $f: \mathbb{Z}_q \to \mathbb{C}$ and $\hat{f}$ is defined by (2.4) then f can be recovered from $\hat{f}$ by the familiar inversion formula

$$f(m) = q^{-1/2} \sum_{n=1}^{q} \hat{f}(n)\, e\left(\frac{nm}{q}\right) .$$

If the support of f is contained in the multiplicative group $G(q)$, with q square free, then f can be recovered _from_ $\hat{f}$ _restricted to_ $G(q)$ by the formula

$$(3.6) \qquad f(m) = q^{-1/2} \sum_{n \in G(q)} \hat{f}(n)W(mn) \ .$$

Applying the Fourier transform to both sides of (3.6), we find that for each integer ℓ,

$$(3.7) \qquad \hat{f}(\ell) = q^{-1/2} \sum_{n \in G(q)} \hat{f}(n)\hat{W}(\ell\tilde{n}) \ ,$$

where $n\tilde{n} \equiv 1 \bmod q$. The formulas (3.6) and (3.7) are special cases of more general identities proved in [1], Theorem 3. The representation (3.4) is established in [1], Theorem 9.

LEMMA 3. _The function_ $f\colon \mathbb{Z}_q \to \mathbb{C}$ _is in the set_ $\mathfrak{G}_q$ _if and only if its Fourier transform_ $\hat{f}$ _satisfies_ (3.7) _and_

$$(3.8) \qquad q^{-1/2}\varphi(q)\mu(q) = \sum_{n \in G(q)} \hat{f}(n) \ .$$

Proof. If f is in the set $\mathfrak{G}_q$ then we have already observed that (3.7) must hold. We may write (2.9) as

$$(3.9) \qquad q^{1/2}\hat{f}(0) = \varphi(q) \ .$$

Then we use (3.7) with $\ell = 0$ and (3.5) to obtain

$$(3.10) \quad \hat{f}(0) = q^{-1/2} \sum_{n \in G(q)} \hat{f}(n)\hat{W}(0) = \mu(q) \sum_{n \in G(q)} \hat{f}(n) \ .$$

The identity (3.8) follows from (3.9) and (3.10).

If $\hat{f}$ satisfies (3.7), we apply the inverse Fourier transform to both sides of (3.7) to deduce (3.6). From [1],

Theorem 9, we see that the support of W is contained in $G(q)$ and so f given by (3.6) must also have its support in $G(q)$. That (3.8) implies (2.9) follows as before and hence we find that f is in the set $\mathfrak{G}_q$.

We now return to consideration of the Hermitian form $Q(f,K)$ given by (3.3). From Lemma 3 it follows that for f in $\mathfrak{G}_q$ we have

$$(3.11) \qquad Q(f,K) = \sum_{\ell=1}^{q} \left| \sum_{n \in G(q)} \hat{f}(n)\hat{W}(\ell\tilde{n}) \right|^2 |\hat{K}(\ell)|^2$$

where $\hat{f}$ restricted to $G(q)$ must also satisfy (3.8). In other words, the problem of minimizing $Q(f,K)$ over all functions f in $\mathfrak{G}_q$ is equivalent to minimizing the right hand side of (3.11) over all functions $\hat{f}:G(q) \to \mathbb{C}$ which satisfy (3.8).

At this point we remark that the problem of minimizing a nonnegative Hermitian form subject to a linear constraint has an elementary solution. Specifically, let H be an $N \times N$ nonsingular Hermitian matrix and $h(\vec{z}) = \vec{z}^* H \vec{z}$ the corresponding form. If $\vec{t} \neq \vec{0}$ is a fixed vector in $\mathbb{C}^N$ then the identity

$$(3.12) \qquad \inf\{h(\vec{z}) : \vec{t}^* \vec{z} = 1\} = (\vec{t}^* H^{-1} \vec{t})^{-1}$$

can be verified readily. Moreover, the problem of minimizing $Q(f,K)$ over f in $\mathfrak{G}_q$ can be viewed as a problem of exactly this type. Thus (3.12) provides a formula for the solution of our extremal problem. However, when this formula is used with $K = K_h$ and either of the representations (3.2) or (3.11), the expression which occurs on the right hand side of (3.12) appears to be very difficult to estimate. While this approach can not be completely discounted, at present it does not seem to be particularly promising.

An alternative method is to apply Cauchy's inequality to the right hand side of (3.11). This produces a diagonal form which majorizes $Q(f,K)$.

THEOREM 4. *Suppose that* q *is square free and for each integer* n *satisfying* $(n,q) = 1$, *let*

$$c(n) = c_K(n) = \sum_{d|q} |\hat{K}(d\bar{d}n)|^2 \varphi(d) . \tag{3.13}$$

Let $R(f,K)$ *be the Hermitian form defined by*

$$R(f,K) = q \sum_{n \in G(q)} c_K(n) |\hat{f}(n)|^2 . \tag{3.14}$$

Then

$$Q(f,K) \leqslant R(f,K) \tag{3.15}$$

for all f *in* $\mathfrak{G}_q$. *If* $f = \chi$ *is the principal Dirichlet character mod* q , *then*

$$Q(\chi,K) = R(\chi,K) = \sum_{n \in G(q)} c_K(n) , \tag{3.16}$$

and

$$\inf\{R(f,K) : f \in \mathfrak{G}_q\} = \varphi(q)\{\varphi(q)^{-1} \sum_{n \in G(q)} (c_K(n))^{-1}\}^{-1} . \tag{3.17}$$

Moreover, the infimum (3.17) *is achieved for at least one function* f *in* $\mathfrak{G}_q$. (*If* $c_K(n) = 0$ *for some* $n \in G(q)$ *the right hand side of* (3.17) *is to be interpreted as zero.*)

Proof. From (3.11) we have

$$Q(f,K) \leqslant \sum_{\ell=1}^{q} |\hat{K}(\ell)|^2 \Big(\sum_{n \in G(q)} |\hat{f}(n)|^2 |\hat{W}(\ell\tilde{n})| \Big)\Big(\sum_{m \in G(q)} |\hat{W}(\ell\tilde{m})| \Big). \tag{3.18}$$

Using (3.5) we obtain

$$\sum_{m \in G(q)} |\hat{W}(\ell\tilde{m})| = q^{1/2} \sum_{d|q} \sum_{\substack{m=1 \\ (m,q)=1 \\ \ell\tilde{m} \equiv d\bar{d} \bmod q}}^{q} 1 \tag{3.19}$$

$$= q^{1/2} \sum_{d|(\ell,q)} \sum_{\substack{m=1 \\ (m,q)=1 \\ \ell \equiv d\bar{d}m \bmod q}}^{q} 1 = q^{1/2}\varphi((\ell,q)) .$$

Now let $\mathcal{D} = \{d\bar{d} : d|q\}$ be the idempotent elements in $\mathbb{Z}_q$. Combining (3.5), (3.18) and (3.19), we find that

$$Q(f,K) \leqslant q^{1/2} \sum_{n \in G(q)} |\hat{f}(n)|^2 \sum_{\ell=1}^{q} |\hat{K}(\ell)|^2 |\hat{W}(\ell\tilde{n})| \varphi((\ell,q))$$

$$= q \sum_{n \in G(q)} |\hat{f}(n)|^2 \sum_{\substack{\ell=1 \\ \ell\tilde{n} \in \mathcal{D}}}^{q} |\hat{K}(\ell)|^2 \varphi((\ell,q))$$

$$= q \sum_{n \in G(q)} |\hat{f}(n)|^2 \sum_{d|q} |\hat{K}(d\bar{d}n)|^2 \varphi(d) .$$

This establishes (3.15).

Since q is square free, we have

$$|\hat{\chi}(n)|^2 = q^{-1} \Big| \sum_{\substack{m=1 \\ (m,q)=1}}^{q} e\Big(\frac{-nm}{q}\Big) \Big|^2 = q^{-1}\varphi((n,q))^2$$

for the principal Dirichlet character. Of course χ is in

$\mathfrak{G}_q$ so we may use (3.3) to obtain

$$(3.20) \qquad Q(\chi,K) = \sum_{n=1}^{q} |\hat{K}(n)|^2 \varphi((n,q))^2 ,$$

and (3.14) to obtain

$$(3.21) \qquad R(\chi,K) = \sum_{n \in G(q)} c_K(n) .$$

Now from the definition (3.13) of $c_K(n)$ we have

$$\begin{aligned}(3.22) \qquad R(\chi,K) &= \sum_{n \in G(q)} \sum_{d|q} |\hat{K}(d\bar{d}n)|^2 \varphi(d) \\ &= \sum_{d|q} \varphi(d) \sum_{\substack{n=1 \\ (n,q)=1}}^{q} |\hat{K}(d\bar{d}n)|^2 \\ &= \sum_{d|q} \varphi(d)^2 \sum_{\substack{m=1 \\ (m,q)=d}}^{q} |\hat{K}(m)|^2 \\ &= \sum_{m=1}^{q} |\hat{K}(m)|^2 \varphi((m,q))^2 \\ &= Q(\chi,K) .\end{aligned}$$

The identities (3.21) and (3.22) prove that (3.16) holds.

Finally, we must minimize $R(f,K)$ over functions $\hat{f}:G(q) \to \mathbb{C}$ satisfying (3.8). If some $c_K(n)$ is zero then an obvious choice of $\hat{f}$ shows that the minimum value is zero and is achieved. Thus, we may assume that each $c_K(n)$ is positive for $(n,q) = 1$. From (3.8) and Cauchy's inequality

$$(3.23)\qquad q^{-1}\varphi(q)^2 = \Big|\sum_{n\in G(q)} \hat{f}(n)\Big|^2$$

$$\leqslant \left(\sum_{n\in G(q)} (c_K(n))^{-1}\right)\left(\sum_{m\in G(q)} c_K(m)\,|\hat{f}(m)|^2\right)$$

and so

$$(3.24)\qquad \varphi(q)^2\left(\sum_{n\in G(q)} (c_K(n))^{-1}\right)^{-1} \leqslant R(f,K) \quad .$$

It is easy to see that there is a case of equality in (3.23) and hence in (3.24). Thus, (3.17) must hold and the infimum is achieved. This completes our proof.

In view of (3.15), the right hand side of (3.17) provides an upper bound for the minimum of $Q(f,K)$ with f in $\mathfrak{G}_q$. By (3.16) this is at least as sharp as the trivial upper bound $Q(\chi,K)$. In fact, the right hand side of (3.17) is equal to $Q(\chi,K)$ only when $c_K(n)$ is constant on $G(q)$. This, of course, is exactly when equality occurs in the inequality between the harmonic and arithmetic mean.

Finally, we may apply Theorem 4 to the inequality (3.1). For square free q and integers h, $1 \leqslant h \leqslant q-1$, we let K_h be defined by (2.11) and note that

$$|\hat{K}_h(n)|^2 = \begin{cases} q^{-1}\left(\dfrac{\sin(\frac{\pi h n}{q})}{\sin(\frac{\pi n}{q})}\right)^2 & \text{if} \quad n \not\equiv 0 \bmod q \\[2ex] 0 \quad \text{if} \quad n \equiv 0 \bmod q \quad . \end{cases}$$

We write $c_h(n) = c_{K_h}(n)$ for the function which is defined by (3.13) with $K = K_h$. By Theorem 4 there exists a function f_h in $\mathfrak{G}_q$ such that

$$Q(f_h,h) \leqslant \varphi(q)\{\varphi(q)^{-1} \sum_{n \in G(q)} (c_h(n))^{-1}\}^{-1} .$$

Applying (3.1) we see that

$$(3.25) \quad T_2(q) \ll 1 + \sum_{h=1}^{M(q)} h^{-2}\{\varphi(q)^{-1} \sum_{n \in G(q)} (c_h(n))^{-1}\}^{-1} .$$

On the other hand, the inequality (1.6) is equivalent to (3.1) with $f_h = \chi$ for each h. Indeed, from (3.16) we may write (1.6) in terms of $c_h(n)$ as

$$(3.26) \quad T_2(q) \ll 1 + \sum_{h=1}^{M(q)} h^{-2}\{\varphi(q)^{-1} \sum_{n \in G(q)} c_h(n)\} .$$

It is clear that (3.25) is sharper than the bound (3.26) but at present we are unable to decide if this improvement is significant. What is required is a method for deciding if the functions $c_h(n)$ have sufficient oscillation to cause their harmonic means to be considerably smaller than their arithmetic means. The effect of such oscillation is evidently rather subtle. We note, for example, that the right hand side of (3.26) is between constant multiples of

$$1 + \sum_{p|q} p^{-1} \log p$$

while the right hand side of (3.25) is at least as large as an absolute constant.

NOTES

1 Research of both authors was supported in part by grant from the National Science Foundation.

2 A proof of the conjectured bound (1.4) for all $\alpha \geqslant 1$ was announced in June, 1984, by H. L. Montgomery and R. C. Vaughan.

REFERENCES

[1] Diamond, H.G., Gerth, F. and Vaaler, J.D., Gauss sums and Fourier analysis on multiplicative subgroups of $\mathbb{Z}_q$, Trans. Amer. Math. Soc., 277(1983), pp. 711-726.

[2] Erdös, P., The difference of consecutive primes, Duke Math. 6(1940), pp. 438 - 441.

[3] ------, Problems and Results in Number Theory, vol. 1, Academic Press (1981).

[4] Hausman, M. and Shapiro, H.N., On the Mean Square Distribution of Primitive Roots of Unity, Com. Pure and Appl. Math., 26(1973), pp. 539 - 547.

[5] Hooley, C., On the difference of consecutive numbers prime to n, Acta Arith. 8(1963), pp. 343 - 347.

[6] ------, On the difference of consecutive numbers prime to n, II, Pub. Math. Debrecen, 12(1965), pp. 39-49.

[7] Norton, K., On Character sums and Power Residues, Trans. Amer. Math. Soc. 167(1972), pp. 203 - 226.

Harold G. Diamond
Department of Mathematics
University of Illinois
Urbana, Ill. 61801

Jeffrey D. Vaaler
Department of Mathematics
University of Texas
Austin, Texas 78712

Some Solved and Unsolved Problems of Mine in Number Theory

Paul Erdös

I.

Let $p_1 < p_2 < \dots$ be the sequence of consecutive primes. Put $p_{n+1} - p_n = d_n$. It is well known that the sequence d_n behaves very irregularly. An old result of Ricci and myself states that the set of limit points of $d_n/\log n$ has positive measure. "It is of course clear to every right thinking person" that the set $d_n/\log n$ is dense in $(0, \infty)$, but we are very far from being able to prove this ("we" here stands for the collective intelligence (or rather stupidity) of the human race). By a result of Westzynthius, ∞ is a limit point of $d_n/\log n$ but no other limit point is known.

Turan and I easily proved that both inequalities $d_n > d_{n+1}$ and $d_n < d_{n+1}$ have infinitely many solutions; we noticed with some annoyance and disappointment that we could not prove that $d_n > d_{n+1} > d_{n+2}$ has infinitely many solutions. We were even more annoyed when we soon noticed that we could not even prove that either $d_n > d_{n+1} > d_{n+2}$ or $d_n < d_{n+1} < d_{n+2}$ has infinitely many solutions. In other words we could not prove that for every n, $(-1)^i(d_{n+i+1} - d_{n+i})$ must have infinitely many changes of sign. I offer 250 dollars for a proof of this conjecture (and 25,000 for a disproof).

Put $f(n) = \log n \log\log n \log\log\log\log n/(\log\log\log n)^2$. Rankin proved in 1938 that

$$\limsup \left(\frac{d_n}{f(n)}\right) > 0 . \tag{1}$$

I offered 10,000 dollars for a proof of

$$\limsup \left(\frac{d_n}{f(n)}\right) = \infty . \tag{2}$$

Thirty years ago I proved that

$$(3) \qquad \limsup \left(\frac{\min(d_n, d_{n+1})}{f(n)} \right) > 0 ,$$

and conjectured that for every k

$$(4) \qquad \limsup \left(\frac{\min(d_n, \ldots, d_{n+k})}{f(n)} \right) > 0 .$$

I could not even prove (4) for $k = 2$. Recently in a brilliant paper Maier proved (4) for every k.

I proved

$$(5) \qquad \liminf d_n/\log n < 1$$

and could not even prove

$$(6) \qquad \liminf \max(d_n, d_{n+1})/\log n < 1 .$$

I offer 500 dollars for a proof of

$$\liminf \max(d_n, d_{n+1}, \ldots, d_{n+k})/\log n < 1$$

and 250 dollars for the proof of (6).

An old conjecture on primes states that there are arbitrarily long arithmetic progressions among the primes. The longest known progression has 17 terms and is due to Weintraub. I conjectured more than 40 years ago that if $a_1 < a_2 < \ldots$ is a sequence of integers for which $\sum_{i=1}^{\infty} \frac{1}{a_i} = \infty$ then the a_i's contain arbitrarily long arithmetic progressions. I offer 3,000 dollars for a proof or disproof. No doubt the following very much stronger result holds: For every k there is an n for which $d_n = \ldots = d_{n+k}$. It is not even known that $d_n = d_{n+1}$ has infinitely many solutions. Rényi and I proved 35 years ago that the density of integers n for which $d_n = d_{n+1}$ is 0, and in fact our proof gives without much difficulty that the density of integers n for which the k numbers $d_n, \ldots, d_{n+k-1}$ are all distinct is one. Let us henceforth consider only those n for which this happens. Order the k integers $d_n, \ldots, d_{n+k-1}$ by size. This gives a permutation $\{i_1, \ldots, i_k\}$ of $1, \ldots, k$. No doubt all the $k!$ permutations occur infinitely often, and in fact if there is justice in heaven or earth each occurs with the same frequence $\frac{1}{k!}$. (There clearly is no justice in heaven or earth, but the conjecture nevertheless holds since there certainly is justice in Mathematics. Unfortunately I cannot at present give a better proof for the conjecture.)

Denote by $F(k)$ the number of permutations which must occur among the $i_1, \ldots, i_k$. $F(2) = 2$ is the simple result of Turán and myself, and perhaps it will not be too difficult to prove $F(3) \geqslant 4$. The following problem which just occurred to me might be of some interest here. Let $a_1 < a_2 < \ldots$ be a sequence of integers for which $a_n/n \log n \to 1$. Put $a_{n+1} - a_n = D_n$ and assume that for every k and almost all n the integers $D_n, \ldots, D_{n+k-1}$ are all distinct. Order the integers $D_n, \ldots, D_{n+k-1}$ by size; this gives a permutation $\{i_1, \ldots, i_k\}$ and denote by $G(k)$ the smallest possible number of these permutations. Again it is easy to see that $G(2) = 2$ and perhaps $G(k)$ can be determined exactly. (The problem is to be understood as follows: We assume only $a_n/n \log n \to 1$ and that for almost all n the k integers $D_n, \ldots, D_{n+k-1}$ are all distinct.) Can one show $F(k) > G(k)$ for some k?

One final problem. De Bruijn, Turán, Katai and I independently investigated the function

$$f(n) = \sum_{p < n} \frac{1}{n-p},$$

hoping to find something new about the primes. We of course noticed that

(7) $$\frac{1}{x}\sum_{n=1}^{x} f(n) \to 1 \quad \text{and} \quad f(n) > c \quad \text{for all} \quad n .$$

The first equation of (7) follows from the prime number theorem and the second from the theorem of Hoheisel. Also $\frac{1}{x}\sum_{n<x} f^2(n) < C$ easily follows from Bruns' method. I once claimed that it follows from Bruns' method that

(8) $$\sum_{n=1}^{x} f^2(n) = x + o(x) .$$

Pomerance noted that I am wrong and (8) (which as far as I know is still open) almost certainly needs more new ideas. Besides the proof of (8) the interesting and difficult questions are:

(9) $$\limsup f(n) = \infty, \quad f(n)/\log\log n \to 0 .$$

Probably

(10) $$\liminf f(n) = 1 ,$$

and perhaps

(11) $$c_1 < \limsup f(n)/\log\log\log n < c_2 , \quad 0 < c_1 \leq c_2 < \infty .$$

I am most doubtful about (11).

I recently conjectured that if $a_1 < a_2 < \ldots$ satisfies $a_n/n \log n \to 1$ and we put $F(n) = \sum_{a_i < n} \frac{1}{n - a_i}$, then 1 is always a limit point of the sequence $F(n)$.

Montgomery found an ingenious and highly nontrivial proof of this conjecture. Ruzsa and I further conjectured that 2 is also a limit point of the sequence $F(n)$. It is a simple exercise to show that no other α, $0 \leq \alpha \leq \infty$ is a compulsory limit point of this sequence.

II.

Now about divisor problems. Denote by $\tau(n)$ the number of divisors of n; $1 = d_1 < d_2 < \ldots < d_{\tau(n)} = n$ are the consecutive divisors of n. I conjectured more than 40 years ago that the density of integers for which

$$\min d_{i+1}/d_i < 2 \tag{12}$$

is 1. More than 30 years ago I proved that the density of the integers satisfying (12) exists but I could never prove that it is 1. At this moment the question is still open and in fact all the recent results of Tenenbaum and myself seem to be consistent with the assumption that the conjecture is false, but nevertheless we both believe it to be true.

Denote by $\tau^*(n)$ the number of integers k such that $2^k < d_i < 2^{k+1}$ for some i. I conjectured that $\tau^*(n)/\tau(n) \to 0$ for almost all n; this would imply conjecture (12). Tenenbaum and I proved that this conjecture is completely wrongheaded and that in fact the density of integers for which $\tau^*(n)/\tau(n) \to 0$ is 0.

Denote by $g(n)$ the number of indices i for which $d_i | d_{i+1}$. At the meeting in Durham in 1979 Montgomery conjectured that for almost all integers $g(n) \neq o(\tau(n))$. I offered Montgomery a bet 10 to 1 that he was wrong. But it turned out that my intuition misled me, and in a forthcoming paper Tenenbaum and I prove Montgomery's conjecture. We also prove that

$$h(n) = \frac{1}{\tau(n)} \sum_{i=1}^{\tau(n)-1} \frac{d_i}{d_{i+1}}$$

has a distribution function. We could not prove that the distribution function is continuous.

Hooley investigated and used the following function:

$$\Delta(n) = \max_t \sum_{t \leq d_i \leq 2t} 1 .$$

Hooley proved

$$(13) \qquad \sum_{n=1}^{x} \Delta(n) < c_0 x (\log x)^c$$

where c is a small positive constant. I proved that

$$(14) \qquad \frac{1}{x} \sum_{n=1}^{x} \Delta(n) \to \infty .$$

Hall and Tenenbaum improved the value of c and also proved

$$(15) \qquad \sum_{n=1}^{x} \Delta(n) > c_1 x \log \log x .$$

Hooley thought that perhaps for every $\varepsilon > 0$ and $x > x_0(\varepsilon)$,

$$(16) \qquad \sum_{n=1}^{x} \Delta(n) < x (\log x)^{\varepsilon}$$

holds. (Note that in the introduction to the paper referred to below, Hooley has qualified this suggestion.) It would be of interest to estimate $\sum_{n=1}^{x} \Delta(n)$ as accurately as possible. It seems likely that a proof of my conjecture $\Delta(n) \geq 2$ for almost all n also will give that $\Delta(n) \to \infty$ if we neglect a sequence of density 0.

Put

$$\Delta_c(n) = \max_t \sum_{t \leq d_i < ct} 1$$

(so that $\Delta(n) = \Delta_2(n)$). Perhaps $\Delta_c(n)/\Delta_2(n) \to 1$ holds for almost all n and every c, $1 < c < \infty$. Also perhaps for every c, $1 < c < \infty$,

$$\sum_{n=1}^{x} \Delta_c(n) / \sum_{n=1}^{x} \Delta_2(n) \to 1 .$$

The following conjecture (the author of which I cannot

place) states: Put ($\mu(n)$ is the well known function of Mobius)

$$r(n) = \max_t \Big| \sum_{1 \leq d_i < t} \mu(d_i) \Big| .$$

Is it true that for almost all n, $r(n) \to \infty$?

R.R. Hall and I investigated

$$(17) \qquad f(n) = \sum_{(d_i, d_{i+1})=1} 1 .$$

We proved that for infinitely many n

$$(18) \qquad f(n) > \exp(\log \log n)^{2-\varepsilon} .$$

During the meeting at Lyon on ordered sets Tenenbaum visited me for a day and we did some "illegal thinking" about $f(n)$. "Illegal thinking" is an important concept introduced by R.L. Graham and myself. It describes the situation when a mathematician works on a problem when he is really supposed to work on another one.

We improved (18) a great deal. Let $n_k = \prod_{j=1}^{k} p_j$ be the product of the first k primes. The prime number theorem implies

$$(19) \qquad n_k = \exp\{(1 + o(1))k \log k\} .$$

Since $\tau(n_k) = 2^k$ we obtain by a simple computation that for at least 2^{k-1} indices i

$$(20) \qquad d_{i+1}/d_i < 1 + \frac{10k \log k}{2^k} .$$

Henceforth we will only consider the d_i satisfying (20). The consecutive divisors satisfying (20) unfortunately do not have to be relatively prime. To overcome this difficulty we construct two consecutive divisors d_i/T, d_x satisfying

(21) $$(d_i/T, d_x) = 1 \quad \text{and} \quad Td_x/d_i \leqslant d_{i+1}/d_i$$

by the following simple process. First of all consider

$$d_i/(d_i, d_{i+1}), \quad d_{i+1}/(d_i, d_{i+1}).$$

These two divisors are certainly relatively prime, but they do not have to be consecutive. Let d_y be the smallest divisor of n_k which is greater than $d_i/(d_i, d_{i+1})$. These two consecutive divisors satisfy the second condition of (21). If the first is not satisfied we repeat the same process with $d_i/(d_i, d_{i+1})$ and d_y and continue. Clearly in a finite number of steps we arrive at two consecutive relatively prime divisors d_z, d_{z+1} which satisfy (21).

The only trouble now is that the same pair d_z, d_{z+1} may originate from many of the consecutive pairs $d_\ell, d_{\ell+1}$. Observe that by (20) and (21) we have

(22) $$d_{z+1}/d_z < 1 + \frac{10k \log k}{2^k}.$$

From (22) and $d_{z+1}/d_z \geqslant 1 + \frac{1}{d_z}$ we immediately obtain

(23) $$d_z \geqslant 2^k/10k \log k.$$

By the prime number theorem all prime factors of n_k -- and therefore of d_z -- are less than $2k \log k$ and thus by (23) ($v(m)$ denotes the number of distinct prime factors of m)

(24) $$v(d_z) > (1 + o(1))(k/\log k).$$

Finally notice that if d_z, d_{z+1} originated by our process from d_i, d_{i+1} we must have $d_z | d_i$. Thus the number of choices of d_i is at most

(25) $$2^{k-v(d_z)} \geqslant 2^{k-(1+o(1))k/\log k}.$$

We remind the reader that the number of possible choices of the pair d_i, d_{i+1} satisfying (20) was at least 2^{k-1}. Thus the number of distinct indices $(d_z, d_{z+1}) = 1$ is by (25) greater than $(\log 2)\exp\{((\log 2) + o(1))\left(\frac{k}{\log k}\right)\}$. Thus finally from (19) we have

$$(26)\qquad \begin{aligned} f(n_k) &\geqslant (\log 2)\exp\{((\log 2) + o(1))\left(\frac{k}{\log k}\right)\} \\ &\geqslant (\log 2)\exp\{((\log 2) + o(1))\left(\frac{\log n_k}{(\log\log n_k)^2}\right)\}. \end{aligned}$$

I hope the reader will agree that our "illegal thinking" was not a complete waste of time.

How close is (26) to being best possible? Clearly

$$f(n) \leqslant \tau(n) \leqslant \exp\{((\log 2) + o(1))\left(\frac{\log n}{\log\log n}\right)\}.$$

Is it true that for every $\varepsilon > 0$ and $n > n_0(\varepsilon)$,

$$(27)\qquad f(n) < \exp(\varepsilon \log n/\log\log n)\ ?$$

At present we cannot decide this question. We also tried to prove that

$$\frac{1}{x \log\log x} \sum_{n=1}^{x} f(n) \to \infty .$$

Here we failed completely. All we could show is that

$$\liminf \frac{1}{x \log\log x} \sum_{n=1}^{x} f(n) > 1 + c$$

for some $1 < c < 2$, which is very unsatisfactory.

Put now $f(n) = f_2(n)$. I tried to estimate $f_3(n)$, the number of indices i for which d_i, d_{i+1}, d_{i+2} are pairwise relatively prime. I never could get a nontrivial result. $f_3(n) > c(\log n/\log\log n)$ holds for infinitely many n, but the proof is essentially trivial and I never could get anything better. (Early in August 1982 Sárközy and I proved that for every r and c there is an n for which $f_r(n) > (\log n)$.)

Put

$$h(n) = \sum_{(d_i\,,\,d_{i+1})=1} (d_i/d_{i+1}) \ .$$

Is it true that for almost all n, $h(n) \to \infty$? Can one get an asymptotic formula -- or a good inequality -- for $\sum_{n=1}^{x} h(n)$?

During my talk at Austin (June 1982) I announced the following fairly recent conjecture of mine. Define

$$L_2(n) = \sum_{i=1}^{\tau(n)-1} (d_{i+1}/d_i - 1)^2 \ . \tag{28}$$

The conjecture states that $L_2(n)$ (and more generally $L_{1+\varepsilon}(n)$) is bounded for infinitely many n. More specifically I conjectured that $L_2(n!)$ is bounded.

Dr. M. Vose proved my first conjecture for $L_{1+\varepsilon}(n)$ in a very ingenious way. His method does not seem to give the boundedness of $L_2(n!)$: This was recently proved by Tenenbaum.

One final remark. A little elementary analysis gives that the boundedness of (28) implies

$$\tau(n) > c(\log n)^2 \ . \tag{29}$$

I expect that (28) implies

$$\tau(n)/(\log n)^2 \to \infty$$

and, in fact, it is quite possible that (28) implies a much sharper lower bound for $\tau(n)$.

III.

Two conjectures from the last century on consecutive integers have been settled in the last decade. Catalan conjectured that 8 and 9 are the only consecutive powers. Tijdeman proved that there is an absolute constant c which can be explicitly determined so that if there are two consecutive powers they must be less than c.

In the first half of the last century it was conjectured that the product of consecutive integers is never a power. After many preliminary results Selfridge and I proved this conjecture. In fact we proved that for every ℓ there is a $p > k$ and $\alpha \not\equiv 0 \pmod{\ell}$ for which $p^{\alpha} \| \prod_{i=1}^{k} (n+i)$. We conjectured that for $k > 3$ there is a $p > k$ for which $p \| \prod_{i=1}^{k} (n+i)$. If true, this seems very deep.

Put ($p(m)$ is the least, $P(m)$ the largest prime factor of m)

$$n + i = a_i^{(n)} b_i^{(n)} \quad \text{where} \quad P(a_i^{(n)}) \leqslant k\,, \quad p(b_i^{(n)}) > k\,.$$

An old conjecture of mine states that

$$\frac{1}{k} \min_{1 \leqslant i \leqslant k} a_i^{(n)} \to 0$$

if k tends to infinity. In other words to every ε and $k > k_0$, we have, for every n,

$$\min_{1 \leqslant i \leqslant k} a_i^{(n)} < \varepsilon k\,. \tag{30}$$

A simple averaging process gives

$$\min_{1 \leqslant i \leqslant k} a_i^{(n)} < Ck$$

for an absolute constant C. I made no progress with the proof of (30). This seems to me to be an attractive conjecture, and I offer 100 dollars for a proof or disproof.

Gordon and I investigated the question of how many consecutive values of $a_i^{(n)}$, $i = 1, 2, \ldots$ can be distinct. Denote by $f(k)$ the largest integer for which there is an integer n so that all the values $a_i^{(n)}$, $1 \leqslant i \leqslant f(k)$ are distinct. We proved $f(k) < (2 + o(1))k$ and conjectured that $f(k) < (1 + o(1))k$. I offer 100 dollars for the proof or disproof of this attractive conjecture. A related problem states: Denote by $h(k)$ the largest integer so that there are at least $h(k)$ distinct numbers among the $a_i^{(n)}$, $1 \leqslant i \leqslant k$. Perhaps there is a constant c so that $h(k) > ck$. I have no real evidence for this conjecture, which very well could be wrong.

Ruzsa and I made some little progress (following an observation of Ruzsa). In (30) εk cannot be replaced by $(\varepsilon k/\log)$ and $f(k) = k + o(k)$ cannot be replaced by $k + 0\left(\frac{k}{\log k}\right)$. I once conjectured that

$$\min_n \sum_{i=1}^{k} \frac{1}{a_i^{(n)}} = h(k)$$

tends to infinity, but our observation shows that $h(k) > c \log \log k$, if true, is certainly best possible. We hope to return to these questions later.

I once conjectured that for every $n \geqslant 2k$, there is an i, $0 \leqslant i < k$, for which $(n-i) \mid \binom{n}{k}$. Schinzel and I proved that this conjecture fails for infinitely many n. Perhaps there is a $c > 0$ so that there is an m, $cn < m \leqslant n$, with $m \mid \binom{n}{k}$.

Schinzel conjectured that for every k there are infinitely many integers n for which $v\left(\binom{n}{k}\right) = k$ ($v(n)$ denotes the number of distinct prime factors of n). The conjecture is of the same depth as the prime k-tuple conjecture of Hardy and Littlewood and is probably hopeless already for $k = 2$.

It is easy to see that for all but a finite number of values of n, $\binom{n}{k}$ has at least $k-1$ distinct prime factors greater than k. Is it true that for every k there are infinitely many values of n for which $v(\binom{n}{k}) = k$ and one of these prime factors is $\leqslant k$? This conjecture is even much more hopeless than that of Schinzel. It almost certainly holds for $k = 2$ but is very doubtful for $k > 2$. In fact denote by $p_i(n, k)$ the i-th largest prime factor of $\binom{n}{k}$. It is not difficult to show that for fixed k and $n \to \infty$, $p_{k-1}(n, k) \to \infty$. I cannot decide if $p_k(n, k) \to \infty$. This is probably false for small values of k.

Selfridge and I investigated the integers $\binom{n}{k}$ for which

$$\binom{n}{k} = p_1 p_2 \cdots p_{k-r}, \quad k < p_1 < \cdots < p_{k-r}.$$

We define r as the deficiency of $\binom{n}{k}$. $\binom{47}{11}$ has deficiency 4 and perhaps this is the largest possible deficiency. Observe that perhaps there are only a finite number of integers n and k with positive deficiency. On the other hand, as far as we know, there may be integers k and n whose deficiency is greater than $k(1-\varepsilon)$, though this seems very unlikely.

R. Graham and I observed that for every $n \geqslant 2k$,

$$(31) \qquad v(\binom{n}{k}) \geqslant (1+o(1))\left(\frac{k}{\log k}\right)\log 4,$$

and that $\log 4$ is best possible. The value of n for which the minimum in (31) is assumed is $(1+o(1))2k$, but we could not prove that for infinitely many k the minimum is really assumed for $n = 2k$.

I conjectured that for every $\varepsilon > 0$ there is a k_0 so that for every $k > k_0$ and any set of k consecutive integers $n+1, \ldots, n+k$ there always is at least one, say

$n+i$, $1 \leqslant i \leqslant k$, which has no divisor d satisfying $\varepsilon k \leqslant d \leqslant k$. Ruzsa observed that the conjecture becomes false if εk is being replaced by $\frac{ck}{\log k}$.

Is it true that there is an absolute constant c so that the number of integers $m+i$, $1 \leqslant i \leqslant n$ which have a prime factor p, $\frac{n}{3} < p < \frac{n}{2}$ is greater than $c(n/\log n)$? I could get nowhere with this easy looking problem and perhaps I overlooked a trivial argument. Observe that for suitable m there may be only one i so that $m+i$ $(1 \leqslant i \leqslant n)$ has a prime factor p satisfying $\frac{n}{2} < p \leqslant n$. To see this choose m so that $m + [\frac{n}{2}]$ is a multiple of all the primes p, $\frac{n}{2} < p \leqslant n$.

To end the paper I state a few more problems.

Is it true that there are infinitely many integers n for which for every $1 \leqslant k < n$ ($[a, b]$ is the least common multiple of a and b)

$$[n+1, \ldots, n+k] > [n-1, \ldots, n-k] \text{ ?} \tag{32}$$

I expect that the answer is affirmative, but that the density of the integers n satisfying (32) is 0. Early in 1983 Straus and I proved the second conjecture.

Put

$$\binom{n}{k} = u(n;k)v(n;k)$$

where $P(u(n;k)) \leqslant k$ and $p(v(n;k)) > k$. A well known theorem of Mahler states that for every $\varepsilon > 0$ there is an $n_0 = n_0(\varepsilon, k)$ so that for every $n > n_0$

$$u(n;k) < n^{1+\varepsilon}. \tag{33}$$

(33) is a very pretty and useful inequality; the only trouble is that it is not effective. An effective inequality

replacing (33) would be useful. It is easy to see that for every η and $k > k_0(\eta)$ there are infinitely many values of n for which

$$u(n;k) > cn \log n \cdot e^{k(1-\eta)} . \tag{34}$$

Perhaps the following inequality holds: There is an absolute constant C so that for every n and k $(n \geq 2k)$

$$u(n;k) < Cn^2 e^{2k} . \tag{35}$$

In view of (34), (35) cannot be too far from being best possible, though perhaps $n^2 e^{2k}$ can be replaced by $n(\log n)^{c_1} e^{k(1+\eta)}$. Perhaps for applications it would be more important to determine the hypothetical constant C explicitly. I could not disprove that $(U(n;k) = \max_{m \leq n} u(m;k))$

$$U(n;k) < C_k n \log n , \tag{36}$$

but perhaps (36) is too optimistic; perhaps (33) holds with $(\log n)^{c_k}$ instead of $\log n$. I could not decide whether

$$\sum_{n=2k}^{\infty} \frac{1}{U(n;k)} \tag{37}$$

diverges. Perhaps this will not be difficult.

Let $k = k(n)$ be the largest integer for which $p(\binom{n}{k}) > k$. I can show that for infinitely many n, $k > \frac{c \log n}{\log \log n}$ but I have no nontrivial upper bound for k.

Denote by $A(u,v)$ the product of those primes p for which $p \,\|\, \prod_{i=1}^{v-u} (u+i)$ ($p \,\|\, d$ means $p \mid d$, $p^2 \nmid d$). Is it true that for almost all squarefree numbers n and all u, v with $u < n < v$, $A(u,v) \geq n$?

REFERENCES

I.

Erdös, P. and A. Rényi, "Some problems and results on consecutive primes," Simon Stevin 27 (1949/50), 115-124.

Rankin, R.A., "The difference between consecutive prime numbers," London Math. Soc. Journal 13 (1938), 242-247.

Erdös, P. and P. Turán, "On some new questions on the distribution of prime numbers," Bull. Amer. Math. Soc. 54 (1948), 371-378. See also P. Erdös, "On the difference of consecutive primes," ibid. 885-889.

Bombieri, E. and H. Davenport, "Small differences between prime numbers," Proc. Roy. Soc. Ser. A 293 (1966), 1-18.

Kátai, I., "A result of consecutive primes," Acta Math. Acad. Sci. Hungar. 27 (1976), 153-159.

II.

Erdös, P., "On the density of some sequences of integers," Bull. Amer. Math. Soc. 54 (1948), 685-692.

------, and R.R. Hall, "On some unconventional problems on the divisors of integers," J. Austral. Math. Soc. (Ser. A) 25 (1978), 479-485.

Hall, R.R. and G. Tenenbaum, "On the average and normal orders of Hooley's Δ-function," J. London Math. Soc. 25 (1982), 392-406.

Hooley, C., "On a new technique and its applications to the theory of numbers," Proc. London Math. Soc. (3) 38 (1979), 115-151.

Erdös, P. and G. Tenenbaum, "Sur la structure de la suite des diviseurs d'un entier," Ann. Inst. Fourier (Grenoble) 31 (1981), 17-37.

III.

Tijdeman, R., "On the equation of Catalan," Acta Arithmetica 29 (1976), 197-209.

Erdös, P. and John Selfridge, "The product of consecutive integers is never a power," Illinois J. Math. 19 (1975), 292-301. This paper contains extensive references to the older literature.

Ecklund, E.F., Jr., R.B. Eggleton, P. Erdös and J.L. Selfridge, "On the prime factorization of binomial coefficients," J. Austral. Math. Soc. 26 (1978), 257-269 and P. Erdös, "On prime factors of binomial coefficients," (in Hungarian), Math. Lapok 28 (1977-1980), 287-296. Both papers contain many references to the older literature.

Erdös, P., "Problems and results on consecutive integers," Publ. Math. Sebrecen, 23 (1976), 271-282, see also: "Problems and results on number theoretic properties of consecutive integers and related questions," Proc. Fifth Manitoba Conference on Numerical Math. Congressus Numerantium XVI. Univ. Manitoba Winnipeg, 1975, 25-44, P. Erdös and E. Straus, "On products of consecutive integers," Number Theory and Algebra, Academic Press 1977, 63-70.

Schinzel, A., "Sur un probleme de P. Erdös," Colloq. Math. 5 (1958), 198-204.

Gaps in Certain Sequences

John B. Friedlander

1. INTRODUCTION

Let $\mathcal{B} = \{b_n | n = 1,2,\ldots\}$ denote a set of positive integers and ϕ a real-valued function satisfying $\phi(y) \to \infty$ as $y \to \infty$. For $X \geqslant 1$ we define $E(X)$ to be the measure of the set of y with $1 \leqslant y \leqslant X$ such that the interval $(y, y+\phi(y)]$ contains no member of $\mathcal{B}$. We consider for certain natural sequences $\mathcal{B}$ the problem of finding functions ϕ growing as slowly as possible, yet such that $E(X) = o(X)$.

For the most part, the sequences we consider possess an asymptotic formula

$$\sum_{\substack{n \in \mathcal{B} \\ n \leqslant X}} 1 \sim \frac{X}{\sigma(X)}$$

where $\sigma(X)$ is also slowly growing, say $\sigma(X) \ll \mathcal{L}^A$ for some A. (We use always the notation $\mathcal{L} = \log X$.) Under such circumstances it seems reasonable to expect that

$$(1) \qquad E(X) = o(X) \qquad \text{for the choice} \quad \phi(X) = g(X)\sigma(X)$$

for <u>any</u> function g satisfying $\lim_{X \to \infty} g(X) = \infty$ arbitrarily slowly. (We always use g in this context.)

It also seems reasonable to expect that (1) is best possible. By this we mean that for any positive constant C, no matter how large, the choice $\phi(X) = C\sigma(X)$ implies that $E(X) = \Omega(X)$.

2. PRIMES

In the case where $\mathcal{B}$ is the set of primes Selberg [12] proved that, under the Riemann hypothesis, one may take $\phi = g\mathcal{L}^2$.

More recently, Heath-Brown [8] has shown that if, in addition, one assumes Montgomery's pair correlation conjecture, then one may take $\phi = g\mathcal{L}$ a result of type (1). This is an example where it seems difficult to show (1) is best possible.

Selberg [12] gave also a method for producing unconditional bounds from zero-density estimates. Harman [6] succeeded in combining this method with the Huxley zero-density estimate and the sifting ideas of Heath-Brown, Iwaniec, and Jutila [9] and has recently (unpublished) shown that one may take $\phi(X) = X^{\frac{1}{12}+\epsilon}$.

3. ALMOST-PRIMES

The corresponding problem with $\mathcal{B}$ a set P_r of almost-primes was first considered by Heath-Brown [7] who showed by sieve methods that one could take $\phi(X) = X^{\frac{1}{11}-\delta}$ (for some small $\delta > 0$) when $\mathcal{B} = P_2$ and $\varphi = \mathcal{L}^{35+\epsilon}$ for $\mathcal{B} = P_3$. Motohashi [11] introduced an analytic method which was sharpened by Wolke [13] and then by Harman [5] who has shown that, for $\mathcal{B} = P_2$, one may take $\phi = \mathcal{L}^{7+\epsilon}$. In contrast with the sieve the latter method has the advantage of producing integers with precisely two prime factors. This breaking of the "parity problem" may conceivably find application to the problem in the case of primes.

The author [2,3] introduced a sieve argument which was not as successful in reducing the number of prime factors but which gave a result of type (1) for the length of the intervals.

For $0 < \alpha < 1$ let $S(\alpha)$ denote the set of positive integers m having no prime factor $< m^{\alpha}$.

(2) Theorem: For any fixed $\alpha < \frac{1}{22}$ we may take

$$\phi = g\mathcal{L} \quad \text{for} \quad \mathcal{B} = S(\alpha) .$$

4. MOMENTS

For the sequence $\mathcal{B} = \{b_n\}$ satisfying $\sum_{b_n \leqslant X} 1 \sim \frac{X}{\sigma(X)}$ one expects the estimate

$$(3) \qquad \sum_{b_n \leqslant X} (b_{n+1} - b_n)^\gamma \ll_{(\gamma)} X(\sigma(X))^{\gamma-1}$$

to hold for each fixed positive γ. Trivially the estimate holds for $\gamma = 0$, $\gamma = 1$ and, by Hölder's inequality, for $0 < \gamma < 1$. The validity of the estimate for any single $\gamma > 1$ is sufficient to imply the result (1) by an elementary argument [2]. Thus, from a result of Hooley [10], one obtains $\phi = g\mathcal{L}^{\frac{1}{2}}$ for the set of integers representable as the sum of two squares, while a result of Erdös [1] gives $\phi = g$ for the square-free integers. We remark that this last example is unusual in that the result (1) is easily seen to be best possible. This fact, which follows from the Chinese remainder theorem, was pointed out to the author by C. Spiro.

The author (unpublished) has modified the methods that give (2) to prove results of type (3) for sets $S(\alpha)$.

<u>Theorem</u>: Let $\mathcal{B} = \{b_n\} = S(\alpha)$ where $0 < \alpha < \frac{1}{25}$. Then for each fixed positive $\gamma < 2$,

$$(4) \qquad \sum_{b_n \leqslant X} (b_{n+1} - b_n)^\gamma \ll X\mathcal{L}^{\gamma-1}$$

while

$$(5) \qquad \sum_{b_n \leqslant X} (b_{n+1} - b_n)^2 \ll X\mathcal{L}^2 .$$

We remark that in the case $\alpha > \frac{1}{2}$, i.e. $\mathcal{B} = \{\text{primes}\}$, Heath-Brown [8] has shown that (5) holds, again assuming the Riemann hypothesis and pair correlation conjectures. His method also gives (4) under the same assumptions. Hölder's inequality shows that, for $\gamma > 1$, (4) is best possible. On the other hand, (5) is almost certainly not.

5. BRUN-TITCHMARSH RESULTS

Results of type (3) for $\gamma > 1$, although they give existence of members of $\mathcal{B}$ in most short intervals, cannot be made to give upper (or lower) bounds of the right order of magnitude. The methods of [2,3] can do this provided we slightly restrict the class of functions g. We denote by h a function with $h(y) \to \infty$ as $y \to \infty$ and such that, for all sufficiently large y,

$$\sup\{h(t)\,|\,y < t \leqslant 2y\} < c \inf\{h(t)\,|\,y < t \leqslant 2y\} \quad .$$

<u>Theorem</u>: Let $h(y)$ be any function as above. There exist $C_i(c) > 0$ such that for all $y \leqslant X$ apart from a set of measure $o(X)$, we have

(6) $$\pi(y + h(y)\log y) - \pi(y) < C_1(c)h(y)$$

and (if $\{b_n\}$ is the set of integers which are a sum of two squares)

(7) $$\sum_{y < b_n \leqslant y + h(y)\log^{\frac{1}{2}} y} 1 < C_2(c)h(y) \quad .$$

Remarks: Here again the lengths are essentially best possible. For sequences of positive density such as the square-free integers such upper bounds follow trivially. The dependence of C_i on c is unimportant since, for example, if h is the r-th iterate of logarithm we may take $c = 1 + \epsilon$ independently of r; the dependence shows up in the $o(X)$.

6. CHEBYSHEV-HOOLEY PROBLEMS

Here we consider briefly the problem of whether most short intervals contain integers with large prime factors. Implicit in Harman's result [5] one has

(8) For fixed $A > 7$ if $\mathcal{B}$ is the set of integers n containing a prime factor $> n\log^{-A} n$ then we may take $\phi = \mathcal{L}^A$.

Using Chebyshev's elementary method [4] one gets only much smaller prime factors but a result of type (1) namely:

(9) If $\mathcal{B}$ is the set of integers n having a prime factor $> \sqrt{n}$ then we may take $\phi = g$.

A modification of the above ideas gives:

(10) If $\alpha > \frac{1}{2}$ and $\mathcal{B}$ is the set of integers n having no prime factor $> n^{\alpha}$, then we may take $\phi = g$.

NOTE added in proof:

The proof of the theorem announced in §4 has now appeared in: Moments of sifted sequences, Math. Ann. 267 (1984), 101-106.

Address:
Scarborough College, University of Toronto
Scarborough, Ontario M1C 1A4

REFERENCES

[1] Erdös, P., Some problems and results in elementary number theory, Publ. Math. (Debrecen) 2 (1951), 103 - 109.

[2-3] Friedlander, J.B., Sifting short intervals, Math. Proc. Cambridge Phil. Soc. 91 (1982), 9-15, II ibid., 92 (1982), 381-384.

[4] ------, Large prime factors in small intervals, Colloq. Math. Soc. Janos Bolyai 34 (to appear).

[5] Harman, G., Almost-primes in short intervals, Math. Ann. 258 (1981), 107-112.

[6] ------, Primes in short intervals, Math. Zeit. 180 (1982), 335-348.

[7] Heath-Brown, D.R., Almost-primes in arithmetic progressions and short intervals, Math. Proc. Cambridge Phil. Soc. 83 (1978), 357-375.

[8] ------, Gaps between primes, and the pair correlation of zeros of the zeta-function, Acta Arith. 41 (1982), 85-99.

[9] ------ and Iwaniec, H., On the difference between consecutive primes, Invent. Math. 55 (1979), 49-69.

[10] Hooley, C., On the intervals between numbers that are sums of two squares, Acta Math. 127 (1971), 279-297.

[11] Motohashi, Y., A note on almost-primes in short intervals, Proc. Japan Acad. Ser. A, 55 (1979), 225-226.

[12] Selberg, A., On the normal density of primes in small intervals and the difference between consecutive primes, Arch. Math. Naturvid. 47 (1943), 87-105.

[13] Wolke, D., Fast-Primzahlen in kurzen Intervallen, Math. Ann. 244 (1979), 233-242.

Prime Numbers and the Pair Correlation of Zeros of the Zeta-Function

D. A. Goldston

The connection between prime numbers and the complex zeros of the zeta function was first discovered by Riemann more than 120 years ago. Of primary importance is the "horizontal" position of the zeros. Riemann conjectured all the complex zeros lie on the line with the real part $1/2$; this is now known as the Riemann Hypothesis (RH). While assuming the Riemann Hypothesis often gives substantially better results for primes than what can be proved at present, it is clear that the finer behavior of primes depends on the "vertical" distribution of zeros. It is only recently that any progress has been made in this direction. In 1972 Montgomery [6] introduced the function

$$(1)\qquad F(x,T) = \sum_{0<\gamma,\gamma'\leq T} x^{i(\gamma-\gamma')} \frac{4}{4+(\gamma-\gamma')^2} \qquad (T \geq 2,\ x \geq 1)\ ,$$

where γ, γ' are the imaginary parts of zeros of $\zeta(s)$. He proved, assuming RH,

$$(2)\qquad F(x,T) = \frac{T\log^2 T}{2\pi x^2}(1+o(1)) + \frac{T}{2\pi}\log x + o(T\log T) \qquad (T\to\infty),$$

uniformly for $1 \leq x \leq T$. For $x \geq T$ the behavior changes, and Montgomery conjectured, on number theoretic grounds,

$$(3)\qquad F(x,T) \sim \frac{T}{2\pi}\log T \qquad (T\to\infty)\ ,$$

uniformly for $T \leq x \leq T^M$, for any number $M \geq 1$. This conjecture and equation (2) imply almost all the zeros of $\zeta(s)$ are simple, and also

$$(4) \qquad \sum_{\substack{0<\gamma,\gamma'\leqslant T\\ 0<\gamma'-\gamma\leqslant u}} 1 = \frac{T}{2\pi}\log T(1+o(1)) \int_0^{\frac{u\log T}{2\pi}} \left(1-\left(\frac{\sin \pi v}{\pi v}\right)^2\right)dv \qquad (T\to\infty)$$

uniformly for $u \cong \frac{2\pi}{\log T}$. ($f \cong g$ means $f \ll g$ and $g \ll f$. When we say an equation holds uniformly for $f \cong g$ we mean it holds uniformly for $0 < \alpha \leqslant f/g \leqslant \beta$ where α and β are any given positive numbers.) The pair correlation function $1 - ((\sin \pi u)/\pi u)^2$ that occurs in (4) has a remarkable interpretation (see [6]).

Gallagher and Mueller [2] used equation (4) to improve some of the classical results on primes proved on the RH. Later Mueller [9], Heath-Brown [5], and the present author [3], [4] used forms of the conjecture (3) and RH to prove results on primes. We will now sketch the method used in [3] and [4].

The starting point is an explicit formula due to Montgomery [6]: for $x \geqslant 1$, and assuming RH,

$$(5) \qquad -2x^{\frac{1}{2}-it}\sum_{\gamma}\frac{x^{i\gamma}}{1+(t-\gamma)^2} = \sum_{n=1}^{\infty}\Lambda(n)a_n(x)n^{-it}$$

$$- \frac{2x^{1-it}}{(1/2+it)(3/2-it)} - x^{-\frac{1}{2}}(\log(|t|+2)+O(1))$$

$$+ O(x^{-3/2}(|t|+2)^{-1}) \quad ,$$

where $a_n(x) = \min((n/x)^{1/2}, (x/n)^{3/2})$. Next, it may be verified directly that

$$F(x,T) = \frac{2}{\pi}\int_{-\infty}^{\infty}\left|\sum_{0<\gamma\leqslant T}\frac{x^{i\gamma}}{1+(t-\gamma)^2}\right|^2 dt \quad ,$$

(therefore $F(x,T) \geqslant 0$), and Montgomery used this to prove

$$F(x,T) = \frac{2}{\pi}\int_0^T \left| \sum_\gamma \frac{x^{i\gamma}}{1+(t-\gamma)^2} \right|^2 dt + O(\log^3 T) \quad . \tag{6}$$

The result (2) is obtained by substituting equation (5) into equation (6) and evaluating the mean value of each term. To obtain information on primes, we use the following method suggested by Montgomery. Consider

$$G_\delta(T) = \left(\frac{\sin \frac{\delta}{2} t}{\frac{\delta}{2} t}\right)\left(\sum_{n=1}^\infty \Lambda(n) a_n(x) n^{-it} - \frac{2x^{1-it}}{(1/2+it)(3/2-it)}\right)$$

where $e^\delta = 1 + \frac{1}{T}$ (so $\delta \sim \frac{1}{T}$), and compute the Fourier transform of $G_\delta(t)$. Here

$$\hat{f}(y) = \int_{-\infty}^\infty f(u) e(-uy)\,du \,, \quad e(u) = e^{2\pi i u} \quad .$$

A calculation gives

$$\hat{G}_\delta(y) = \frac{2\pi}{\delta} \sum_{|y + \frac{\log n}{2\pi}| < \delta/4\pi} \Lambda(n) a_n(x) - \frac{2\pi x}{\delta} \int_{xe^{-\delta/2}}^{xe^{\delta/2}} a_v(e^{-2\pi y}) \frac{dv}{v} \,.$$

Consequently, applying Parseval's formula with (5) and simplifying, we obtain, for $x \geqslant 1$ and assuming RH,

$$\int_0^\infty \left(\sum_{y<n<y+\frac{y}{T}} \Lambda(n) a_n(x) - x \int_{xe^{-\delta}}^x a_v(y) \frac{dv}{v} \right)^2 \frac{dy}{y} \tag{7}$$

$$= \frac{4x\delta^2}{\pi} \int_0^\infty \left(\frac{\sin \frac{\delta}{2} t}{\frac{\delta}{2} t}\right)^2 \left| \sum_\gamma \frac{x^{i\gamma}}{1+(t-\gamma)^2} \right|^2 dt + O\left(\frac{\log^2 T}{T}\right) .$$

We shall denote this equation by

$$L(x,T) = R(x,T) \quad . \tag{8}$$

By equation (6) we see $R(x,T)$ is essentially an average in T of $F(x,T)$, and therefore it is routine to evaluate $R(x,T)$ if we know something about $F(x,T)$. We find, for $x \geqslant T$,

$$R(x,T) \begin{cases} \ll \frac{x}{T}\log^2 T & \text{trivially} \\ \ll \frac{x}{T}\log T & \text{if } F(x,T) \ll T\log T \\ \sim \frac{x}{T}\log T & \text{if } F(x,T) \sim \frac{T}{2\pi}\log T \end{cases} \quad . \tag{9}$$

Here the assumption on $F(x,T)$ may be localized; for example if $F(x,t) \sim \frac{t}{2\pi}\log t$ for $T_0 \leqslant t \leqslant T_1 \log T_1$ where $T_0/T \to 0$ and $T_1/T \to \infty$, then $R(x,T) \sim \frac{x}{T}\log T$ (for the same x).

Now consider $L(x,T)$. In intervals without primes (and prime powers) the sum in $L(x,T)$ vanishes and we can evaluate the integral there. Consequently $L(x,T)$ provides an upper bound for the number of long intervals without primes. We find, with $T = 3x/H$,

$$L(x,T) \gg \frac{x}{T^2} \sum_{\substack{\frac{x}{2} \leqslant p_n \leqslant x \\ p_{n+1} - p_n \geqslant H}} (p_{n+1} - p_n) \tag{10}$$

and hence, combining (9) and (10) and adding the results for $x, \frac{x}{2}, \frac{x}{4}, \ldots,$ we conclude

$$\sum_{\substack{p_n \leqslant x \\ p_{n+1} - p_n \geqslant H}} (p_{n+1} - p_n) \ll \frac{x}{H}\log x \tag{11}$$

subject to RH and $F(x,T) \ll T \log T$. Taking the last term in the sum only, we conclude

(12) $$p_{n+1} - p_n \ll \sqrt{p_n \log p_n} \ .$$

These results may be compared with the bounds on the RH alone of $\frac{x}{H} \log^2 x$ in (11) and $\sqrt{p_n} \log p_n$ in (12) due to Selberg and Cramér, respectively. Using a different argument, Heath-Brown and I have proved

(13) $$p_{n+1} - p_n = o(\sqrt{p_n \log p_n})$$

assuming RH and $F(x,T) \sim \frac{T}{2\pi} \log T$ for $T^{2-\epsilon} \leqslant x \leqslant T^{2+\epsilon}$, $\epsilon > 0$.

In order to deal with small differences between primes, $L(x,T)$ must be dealt with more carefully. We note that if y/T is not too large, then in the expression

$$\sum_{y < n < y + \frac{y}{T}} \Lambda(n) a_n(x)$$

we may replace $a_n(x)$ by $a_y(x)$ with a small error. We then obtain (see [4]), for $x \geqslant T$,

(14) $$L(x,T) = 4x^3 \int_x^\infty H(y) \frac{dy}{y^5} + O((\frac{x}{T})^2 \frac{\log^2 x}{T}) \ ,$$

where

$$H(x) = \int_0^x (\psi(y + \frac{y}{T}) - \psi(y) - \frac{y}{T})^2 dy \ , \ \psi(x) = \sum_{n \leqslant x} \Lambda(n) \ .$$

Combining this result with (9) and using a differencing argument, we obtain, on RH and $F(x,T) \sim \frac{T}{2\pi} \log T$,

(15) $$H(x) \sim \frac{x^2 \log T}{2T} \qquad (T \leqslant x \leqslant T^{2-\epsilon}) \quad .$$

From (15) we obtain, using another breaking up argument,

(16) $$J(X) = \int_X^{2X} (\pi(x+h) - \pi(x))^2 dx \sim (\lambda + \lambda^2)X \qquad (X \to \infty)$$

where $h \sim \lambda \log X$, λ any fixed positive number. We have used the RH and the conjecture $F(x,T) \sim \frac{T}{2\pi} \log T$ uniformly for $x \cong T \log T$ in proving (16). This result agrees with the conjecture that the primes are distributed around their average in a Poisson distribution.

As a consequence, we obtain a result of Heath-Brown; namely

(17) $$\liminf_{n \to \infty} \left(\frac{p_{n+1} - p_n}{\log p_n} \right) = 0 \quad ,$$

subject to the same conjectures. For if not, the lim inf = $\mu > 0$ and for $0 < \lambda < \mu$ and x sufficiently large we have $(\pi(x+\lambda \log x) - \pi(x))^2 = \pi(x+\lambda \log x) - \pi(x)$, and thus by the prime number theorem

(18) $$J(X) \sim \lambda X \qquad (x \to \infty)$$

contrary to (16).

We conclude by mentioning a connection between these last results and additive prime number theory. Let

(19) $$Z(X;2n) = \sum_{\substack{X \leqslant p, p' \leqslant 2X \\ p' - p = 2n}} (\log p)(\log p')$$

and

(20) $$T(\alpha) = \sum_{n=-K}^{K} t(n) e(2n\alpha) \geqslant 0 \quad ,$$

where the $t(n)$ are real and $t(-n) = t(n)$. Bombieri and Davenport [1] proved, for $K \leqslant (\log X)^C$, any fixed C, and any $\epsilon > 0$ and $X \geqslant X_o(\epsilon)$,

$$\sum_{n=1}^{K} t(n)Z(X;2n) > 2X \sum_{n=1}^{K} t(n)H(n) - (\tfrac{1}{4}+\epsilon)t(0)\, X \log X \tag{21}$$

where

$$H(n) = H \prod_{\substack{p|n \\ p>2}} \frac{p-1}{p-2}, \qquad H = \prod_{p>2} (1-(p-1)^{-2}) \ .$$

Now let

$$I(X) = I(X,h) = \int_X^{2X} (\psi(x+h) - \psi(x) - h)^2 dx \ . \tag{22}$$

On multiplying out, we find, using the prime number theorem,

$$I(X) \geqslant hX(\log X + O(1)) + 4 \sum_{1 \leqslant n \leqslant h/2} t(n)Z(X,2n) - h^2X - |E| \ , \tag{23}$$

where $t(n) = [\frac{h}{2}] - n$, and $|E| \ll h^3 \log^2 X + h^2 X(\log X)^{-C}$. With this choice of $t(n)$, we have

$$T(\alpha) = \left(\frac{\sin(2\pi\,[h/2]\,\alpha)}{\sin(2\pi\alpha)}\right)^2 \geqslant 0 \ .$$

Hence, using (21) in (23) together with the fact

$$\sum_{1 \leqslant n \leqslant h/2} ([h/2]-n)H(n) = \sum_{1 \leqslant \nu \leqslant [h/2]-1} \left(\sum_{n=1}^{\nu} H(n)\right)$$

$$= \sum_{1 \leqslant \nu \leqslant [h/2]-1} (\nu + O(\log \nu))$$

(see [7] p.149)

$$\geqslant \frac{h^2}{8} + O(h \log h) \ ;$$

we conclude, for $1 \leq h \leq (\log X)^C$,

(24) $$I(X) \geq (\tfrac{1}{2} - \epsilon)hX \log X \qquad (X \geq X_o(\epsilon)) \quad .$$

Taking $h = \lambda \log X$, we obtain an unconditional lower bound for $J(X)$:

(25) $$J(X) \geq ((\tfrac{1}{2} - \epsilon)\lambda + \lambda^2)X \qquad (X \geq X_o(\epsilon)) \quad .$$

This may be compared with the trivial result

(26) $$J(X) \geq (\max(\lambda,\lambda^2) - \epsilon)X \qquad (X \geq X_o(\epsilon)) \quad .$$

Comparing (25) and (18), we see (18) can not hold for $\lambda > \frac{1}{2} + \epsilon$ and hence unconditionally

(27) $$\liminf_{n \to \infty} \frac{p_{n+1} - p_n}{\log p_n} \leq 1/2 \quad .$$

This was obtained by Bombieri and Davenport from (21) directly; incorporating a sieve upper bound in (21) they improve on this a little.

Finally, we mention the above method can be extended conditionally for larger h. The result obtained is, assuming GRH, and letting $h = X^\alpha$,

(28) $$I(X) \geq (\tfrac{1}{2} - 3\alpha - \epsilon)hX \log X \qquad (X \geq X_o(\epsilon))$$

uniformly for $0 \leq \alpha \leq 1/6$. This may be compared with Mueller's result [8]; assuming RH and a strong form of (4), $h = X^\alpha$,

(29) $$I(X) \sim (1 - \alpha)hX \log X \qquad (X \to \infty) \quad ,$$

uniformly for $0 \leq \alpha \leq 1 - \epsilon$.

REFERENCES

1 Bombieri, E. and Davenport H., Small differences between prime numbers, Proc. Roy. Soc. Ser. A, 293 (1966), 1-18.

2 Gallagher, P.X. and Mueller, J., Primes and zeros in short intervals, J. reine angew. Math, 303 (1978), 205-220.

3 Goldston, D.A., Large differences between consecutive prime numbers, Thesis, U.C. Berkeley 1981.

4 ------, The second moment for prime numbers,(unpublished)

5 Heath-Brown, D.R., Gaps between primes, and the pair correlation of zeros of the zeta-function, Acta Arith., 41 (1982), 85 - 99.

6 Montgomery, H.L., The pair correlation of zeros of the zeta function, Proc. Symp. Pure Math. 24, A.M.S. Providence (1973), 181-193.

7 ------, Topics in multiplicative number theory, Lecture Notes in Mathematics, Vol. 227, Springer-Verlag, Berlin (1971).

8 Mueller, J., Primes and zeros in short intervals, Thesis, Columbia University, 1976.

9 ------, On the difference between consecutive primes, Recent Progress in Analytic Number Theory, Vol. 1, Academic Press (1981), 269-273.

Institute for Advanced Study
Princeton, New Jersey 08540
Current Address:
Department of Mathematics
San José State University
San José, California 95192

NOTE added in proof:

The proof of (13) has appeared in: D.R. Heath-Brown and D.A. Goldston, "A note on the differences between consecutive primes," Math. Ann. 266 (1984), 317 - 320.

The proof of (25) and (28) has appeared in: D.A. Goldston, "The second moment for prime numbers," Quart. J. Oxford (2), 35 (1984), 153 - 163.

In joint work with H.L. Montgomery, the results in this paper have been extended considerably. We can now prove that, assuming the Riemann Hypothesis, equations (3) and (29) are equivalent statements. This and other related results will appear in a paper to appear in the Proceedings of the 1984 Stillwater Conference on Analytic Number Theory and Diophantine Problems (to be published by Birkhauser).

A Formula of Landau and Mean Values of $\zeta(s)$

S. M. Gonek

Let $\rho = \beta + i\gamma$ denote a complex zero of the Riemann zeta function, $\zeta(s)$. A remarkable formula of Landau [2] (also see Titchmarsh [4;pp. 61-62]) states that for fixed $x > 1$ and $T \to \infty$

$$(1) \qquad \sum_{0<\gamma\leq T} x^{\rho} = -\frac{T}{2\pi}\Lambda(x) + O(\log T) \quad ,$$

where $\Lambda(x) = \log p$ if $x = p^k$ for some prime p and positive integer k, and $\Lambda(x) = 0$ for all other real x. This can be proved by estimating the integral

$$(2) \qquad \frac{1}{2\pi i} \int_{\mathcal{R}} \frac{\zeta'}{\zeta}(s)\, x^s \, ds \quad ,$$

where $\mathcal{R}$ is a suitably chosen rectangle enclosing those zeros ρ for which $0 < \gamma \leq T$.

Striking as (1) is, it has little utility because it is not uniform in x. It is possible, however, by keeping the estimates of (2) explicit in x and T, to prove the following uniform version of (1).

THEOREM 1. Let $x, T > 1$. Then

$$(3) \qquad \sum_{0<\gamma\leq T} x^{\rho} = -\frac{T}{2\pi}\Lambda(x) + O(x \log 2x \log\log 3x)$$

$$+ O(x \log 2T) + O(\log x \min(T, \frac{x}{\langle x \rangle}))$$

$$+ O(\min(\frac{\log T}{\log x}, T \log T)) \quad ;$$

here $\langle x \rangle$ is the distance from x to the nearest prime power other than x.

Note that a trivial estimate for our sum is $\ll x\, T \log T$ ($\sqrt{x}\, T \log T$ on the Riemann hypothesis) since there are $\sim \frac{T}{2\pi} \log T$ zeros with $0 < \gamma \leq T$. The large number of error terms in (3) reflects the varied behavior of the sum in different x ranges. Thus, the last error term is significant when x is near 1, the next-to-last when x is near a prime power. Finally, we observe that (3) has a particularly simple form if x is an integer such that $2 \leq x \ll T$, namely

$$\sum_{0<\gamma\leq T} x^{\rho} = -\frac{T}{2\pi}\Lambda(x) + O(x \log 2T \operatorname{loglog} 3T) .$$

Theorem 1 may be used to estimate various sums involving zeros of the zeta function. For example, one may use it to prove

THEOREM 2. Assume the Riemann hypothesis. Let T be large, $L = \frac{1}{2\pi}\log T$, and $|\alpha| \leq L$ with α real. Then

$$(4) \qquad \sum_{0<\gamma\leq T} |\zeta(1/2 + i(\gamma+\alpha/L))|^2$$

$$= \left(1 - \left(\frac{\sin \pi\alpha}{\pi\alpha}\right)^2\right) \frac{T}{2\pi}\log^2 T + O(T \log^{7/4} T) .$$

The constant implied by O is absolute.

In [1] we gave asymptotic formulae for the sums

$$\sum_{0<\gamma\leq T} \zeta^{(\mu)}(\rho + i\alpha/L)\zeta^{(\nu)}(1-\rho-i\alpha/L) \qquad (\mu,\nu = 0,1,\ldots) .$$

where $\zeta^{(\mu)}(s)$ is the μth derivative of $\zeta(s) = \zeta^{(0)}(s)$. Theorem 2 is the most interesting special case of these formulae. In fact, from (4) J. Mueller [3] deduced that

$$\lambda = \lim_n \sup(\gamma_{n+1} - \gamma_n)\frac{\log \gamma_n}{2\pi} > 1.9 \quad ,$$

where $0 < \gamma_1 \leqslant \gamma_2 \leqslant \ldots$ are the ordinates of the zeros of $\zeta(s)$ in the upper half-plane. Previously it was only known that $\lambda > 1$.

We now use Theorem 1 to sketch a proof of Theorem 2 which is much shorter than that given in [1]; detailed proofs of both theorems will appear elsewhere. We use the notation $A \approx B$ below to mean that $A = B +$ error terms.

We begin with the approximate functional equation for $\zeta(s)$ (see Titchmarsh [5;p.69]) from which it follows that

$$\zeta(\tfrac{1}{2} + i(\gamma + \alpha/L)) = \sum_{n \leqslant x} n^{-\frac{1}{2} - i(\gamma + \alpha/L)} + O(\log^{\frac{1}{4}} \gamma) \, ,$$

where $x = x(\gamma) = \gamma/2\pi\sqrt{\log \gamma}$.

Our problem is essentially to show that

$$\sum_{0 < \gamma \leqslant T} \left| \sum_{n \leqslant x} n^{-\frac{1}{2} - i(\gamma + \alpha/L)} \right|^2 \approx \left(1 - \left(\frac{\sin \pi\alpha}{\pi\alpha}\right)^2\right) \frac{T}{2\pi} \log^2 T \, .$$

In order to avoid minor difficulties, however, we will show instead that

$$B = \sum_{0 < \gamma \leqslant T} \left| \sum_{n \leqslant X} n^{-\frac{1}{2} - (\gamma + \alpha/L)} \right|^2 \approx \left(1 - \left(\frac{\sin \pi\alpha}{\pi\alpha}\right)^2\right) \frac{T}{2\pi} \log^2 T \, ,$$

where now $X = T/2\pi\sqrt{\log T}$ is dependent of γ.
Squaring out and changing the order of summation, we have

$$B = \sum_{m,n \leqslant X} \frac{1}{\sqrt{mn}} \sum_{0 < \gamma \leqslant T} \left(\frac{n}{m}\right)^{i(\gamma + \alpha/L)} \, .$$

The terms in (m,n) and (n,m) are conjugate, so

$$B = \sum_{n\leqslant X}\frac{1}{n}\sum_{0<\gamma\leqslant T} 1 + 2\,\mathcal{R}e \sum_{m<n\leqslant X}\frac{1}{n}\left(\frac{n}{m}\right)^{i\alpha/L}\sum_{0<\gamma\leqslant T}\left(\frac{n}{m}\right)^{\frac{1}{2}+i\gamma}$$

(5)

$$= B_1 + 2\,\mathcal{R}e\, B_2 ,$$

say. Since

$$\sum_{n\leqslant X}\frac{1}{n} \sim \log X \sim \log T \quad ,$$

and the number of $\gamma \in (0,T]$ is $\sim \frac{T}{2\pi}\log T$,

(6) $$B_1 \sim \frac{T}{2\pi}\log^2 T \quad .$$

Now by the Riemann hypothesis and Theorem 1, the innermost sum in B_2 equals $-\frac{T}{2\pi}\Lambda(\frac{n}{m})$ plus error terms. Hence

$$B_2 \approx -\frac{T}{2\pi}\sum_{m<n\leqslant X}\frac{\Lambda(\frac{n}{m})}{n}\left(\frac{n}{m}\right)^{i\alpha/L} \quad .$$

The term in (m,n) vanishes if $m \nmid n$, so we may write

$$B_2 \approx -\frac{T}{2\pi}\sum_{km\leqslant X}\frac{\Lambda(k)}{k^{1-i\alpha/L}m}$$

$$\approx -\frac{T}{2\pi}\sum_{k\leqslant X}\frac{\Lambda(k)}{k^{1-i\alpha/L}}\sum_{m\leqslant X/k}\frac{1}{m}$$

$$\approx -\frac{T}{2\pi}\sum_{k\leqslant X}\frac{\Lambda(k)}{k^{1-i\alpha/L}}\log X/k \quad .$$

The last sum equals

$$\int_1^X \frac{\log X/u}{u^{1-i\alpha/L}}\, d\,\Psi(u) \quad ,$$

where $\Psi(u) = \sum_{n<u} \Lambda(n)$, and by the prime number theorem with remainder term, this is essentially

$$\int_1^X \frac{\log X/u}{u^{1-i\alpha/L}}\, du = \frac{1 + i\alpha/L \log X - X^{i\alpha/L}}{(\alpha/L)^2} .$$

Therefore,

$$2\,\mathcal{R}e\, B_2 \approx -\frac{T}{\pi} \frac{1 - \cos(\alpha/L \log X)}{(\alpha/L)^2}$$

$$\approx -\frac{T}{2\pi}\left(\frac{\sin(\alpha/2L \log X)}{\alpha/2L}\right)^2 ,$$

or, since $\log X \sim \log T$ and $L = \frac{1}{2\pi} \log T$,

$$2\,\mathcal{R}e\, B_2 \approx -\frac{T}{2\pi}\log^2 T \left(\frac{\sin \pi\alpha}{\pi\alpha}\right)^2$$

Combining this with (5) and (6), we obtain

$$B \approx \left(1 - \left(\frac{\sin \pi\alpha}{\pi\alpha}\right)^2\right) \frac{T}{2\pi} \log^2 T ,$$

as desired.

REFERENCES

[1] Gonek, S.M., 'Mean values of the Riemann zeta-function and its derivatives', Inventiones Math. 75(1984), 123-141.

[2] Landau, E., 'Uber die Nullstellen der Zetafunction', Math. Annalen 71 (1911), 548-564.

[3] Mueller, J., 'On the difference between consecutive zeros of the Riemann zeta function', J. Number Theory 14 (1982), 327-331.

[4] Titchmarsh, E.C., _The zeta-function of Riemann_, Hafner Publishing Co., New York, 1972.

[5] ------, _The theory of the Riemann zeta-function_, Oxford University Press, 1951.

Department of Mathematics
Univeristy of Rochester
Rochester, N.Y. 14627

Large Values of the Riemann Zeta-Function

S. W. Graham

1. Introduction. Let $s = \sigma + it$ be a complex variable and let $\zeta(s)$ denote Riemann's zeta function. Here we are concerned with the problem of obtaining upper bounds for the number of times $\zeta(s)$ assumes large values in the half-plane $\sigma \geqslant 1/2$.

Van der Corput ([7], Theorems 5.12 and 5.13) proved that if $R_q = 2^{q+2} - 2$ and $\sigma_q = 1 - (q+2)/R_q$ then

(1.1) $$\zeta(\sigma_q + it) \ll t^{1/R_q} \log t \qquad (t \geqslant 2) .$$

In particular, $q = 1$ yields

(1.2) $$\zeta(1/2 + it) \ll t^{1/6} \log t \qquad (t \geqslant 2)$$

and $q = 2$ yields

(1.3) $$\zeta(5/7 + it) \ll t^{1/14} \log t \qquad (t \geqslant 2) .$$

The exponents in (1.1) have been improved slightly; see Kolesnik [3] and Phillips [6].

Heath-Brown [2] combined van der Corput's method with Halász's large values method to prove

THEOREM A. *Suppose* $t_1, t_2, \ldots, t_R$ *are real numbers such that*

$$T < t_m \leqslant 2T , \quad |t_m - t_n| \geqslant 1 \quad (m \neq n)$$

and

$$|\zeta(1/2+it_m)| \geqslant V .$$

Then $R \ll T^2V^{-12}\log^{16}T$.

Note that Theorem A contains (1.2) except for an unimportant factor of $\log^{1/3}T$. Another striking consequence of Heath-Brown's result is the estimate

$$\int_0^T |\zeta(1/2+it)|^{12}dt \ll T^2\log^{17}T .$$

The purpose of this paper is to extend Heath-Brown's techniques to values of $\sigma > 1/2$. To avoid overwhelming technical complications, we shall concentrate on the case $\sigma = 5/7$. We prove

THEOREM 1. *Let* $t_1,\ldots,t_R$ *be real numbers such that*

$$T < t_m \leqslant 2T , \quad |t_m - t_n| \geqslant 2\pi \quad (m \neq n)$$

and

$$|\zeta(5/7+it_m)| \geqslant V .$$

Then

$$R \ll T^{14}V^{-196}\log^{424}T .$$

Note that Theorem 1 contains (1.3) except for a power of log T. Another consequence of Theorem 1 is the

COROLLARY. $\int_0^T |\zeta(5/7+it)|^{196}dt \ll T^{14}\log^{425}T$.

Heath-Brown's proof of Theorem A appeals to an explicit formula, due to Atkinson [1], for

$$\int_0^T |\zeta(1/2+it)|^2 dt \, .$$

No such formula for the line $\sigma = 5/7$ is available in the literature. I have instead adopted an alternative approach that is inspired by a comment in Heath-Brown's paper. The heart of this approach is embodied in Lemma 4, which is a slight variation on a result of Gallagher (see [4], Lemma 1.9). As Heath-Brown remarks, this may be regarded as an analog of Weyl's inequality for exponential sums. The estimate (1.3) requires two applications of Weyl's inequality. Similarly, Theorem 1 requires two applications of Lemma 4. Although we shall not give the details, the method developed here allows one to prove Theorem A without recourse to Atkinson's formula. Estimate (1.2) requires one application of Weyl's inequality; the proof of Theorem A requires one application of Lemma 4.

I believe that the method presented here could be adapted to prove

$$\int_0^T |\zeta(\sigma_q + it)|^{14R_q} dt \ll T^{14} (\log T)^{A(q)} \tag{1.4}$$

where σ_q and R_q are as in (1.1) and $A(q)$ is some positive function of q. Note that when $q = 1$, (1.4) states that

$$\int_0^T |\zeta(1/2+it)|^{84} dt \ll T^{14} \log^A T \, .$$

This is weaker than Theorem A, and it reflects the fact that there is one aspect of Heath-Brown's argument that I have been unable to take advantage of here. This point will be discussed further is Section 8.

I am happy to record my gratitude to Roger Heath-Brown for several useful discussions.

2. Notation. We follow standard number theoretic notation in defining $e(x)$ to be $e^{2\pi i x}$ and $\mathcal{L}$ to be $\log T$. We use $f \asymp g$ to mean that $f \ll g$ and $g \ll f$. All constants implied by "O", "$\ll$", and "$\asymp$" are absolute.

We use the method of exponent pairs as developed by Phillips [6]. We do not make full use of the power of this method, for we appeal only the simple exponent pairs $(0,1)$, $(1/6,2/3) = AB(0,1)$, and $(1/14,11/14) = A^2B(0,1)$. With a little more work, we could have avoided any mention of exponent pairs and instead appealed to Theorem 5.13 of Titchmarsh [7]. But we prefer not to eschew the convenient language of exponent pairs.

For the convenience of the reader, we give here a list of the important parameters that appear in Sections 4 through 6. We also give their values in terms of the basic parameters T,N,V and W.

$$G \asymp V^2N^{2\sigma-1}\mathcal{L}^{-2} .$$

$$H \asymp V^{-2}N^{2-2\sigma}\mathcal{L}^{2} .$$

$$J \asymp W^2V^6N^{6\sigma-4}\mathcal{L}^{-6} .$$

$$K \asymp W^{-2}V^{-4}N^{4-4\sigma}\mathcal{L}^{4} .$$

$$L \asymp TW^{-2}V^{-6}N^{3-6\sigma}\mathcal{L}^{6} .$$

$$M \asymp TW^{-6}V^{-14}N^{10-14\sigma}\mathcal{L}^{14} .$$

3. Smoothing lemmas. We use several smoothing arguments to facilitate the treatment of exponential sums; the details are developed in this section.

LEMMA 1. Let

$$g(x) = \int_{-\infty}^{\infty} \left(\frac{\sin \pi t}{\pi t}\right)^4 e(xt)dt .$$

Then

(i) supp $g \subseteq [-2,2]$,

(ii) $\sum_n g(x+n) = 1$ for all real x, and

(iii) g has two continuous derivatives.

Proof. It is well known that

$$g_1(x) = \int_{-\infty}^{\infty} \left(\frac{\sin \pi t}{\pi t}\right) e(tx)dt = \begin{cases} 1 & \text{if } |x| < 1/2 , \\ 1/2 & \text{if } |x| = 1/2 , \\ 0 & \text{if } |x| > 1/2 . \end{cases}$$

Since g is the 4-fold convolution of g_1 with itself, (i) follows. By the Fourier inversion formula

$$\hat{g}(t) = \left(\frac{\sin \pi t}{\pi t}\right)^4 ,$$

and by the Poisson summation formula,

$$\sum_n g(x+n) = \sum_m \hat{g}(m)e(mx) = 1 ,$$

This proves (ii). To prove (iii), we note that if $r \leqslant 2$ then

$$\left(\frac{d}{dx}\right)^r g(x) = \int_{-\infty}^{\infty} \left(\frac{\sin \pi t}{\pi t}\right)^4 (2\pi it)^r e(tx)dt .$$

The right hand side converges since the integrand is majorized by $(1+|t|)^{-2}$. It is continuous since it is the Fourier transform of an L^1 function.

LEMMA 2. Suppose $1/2 \leqslant \sigma \leqslant 1$, $t \geqslant 1$, and $T < t \leqslant 2T$, and let $\mathcal{N} = \{N = e^r : 0 \leqslant r \leqslant \mathcal{L} + 2\}$. Then

$$\zeta(\sigma - 2\pi it) = \sum_{N \in \mathcal{N}} \sum_{n} g\left(\log(\tfrac{n}{N})\right) n^{-\sigma} e(t \log n) + O(1) .$$

Proof. We start from the easily proved result

$$\zeta(\sigma - 2\pi it) = \sum_{n \leqslant T} n^{-\sigma} e(t \log n) + O(1) .$$

(See Titchmarsh [7], Theorem 4.11.) Applying (ii) of Lemma 1, we find that

$$\sum_{n \leqslant T} n^{-\sigma} e(t \log n) = \sum_{n \leqslant T} n^{-\sigma} e(t \log n) \sum_{r} g(-r + \log n)$$

$$= \sum_{r} \sum_{1 \leqslant n \leqslant T} n^{-\sigma} e(t \log n) g(-r + \log n)$$

Now $g(-r + \log n) = 0$ unless $e^{r-2} \leqslant n \leqslant e^{r+2}$. Consequently

$$\sum_{n} e(t \log n) n^{-\sigma} = \sum_{N \in \mathcal{N}} \sum_{\substack{Ne^{-2} \leqslant n \leqslant Ne^{2} \\ 1 \leqslant n \leqslant T}} g\left(\log(\tfrac{n}{N})\right) n^{-\sigma} e(t \log n) .$$

The condition $1 \leqslant n \leqslant T$ in the second sum is vacuous unless $Ne^2 \geqslant T$. These terms contribute

$$\ll T^{-\sigma} \max_{T \leqslant M \leqslant Te^2} \left| \sum_{T < n \leqslant M} e(t \log n) \right| .$$

By Theorem 5.9 of Titchmarsh [7], this is

$$\ll T^{1/2-\sigma} \ll 1 .$$

We conclude this section by giving two lemmas that are basically due to Gallagher. Lemma 3 is taken directly from Montgomery [4], Lemma 1.1. Lemma 4 is Lemma 1.9 of Montgomery [4], but we have introduced an extra smoothing factor.

LEMMA 3. *Suppose* f *is a continuous real-valued function on* $[a,b]$. *Then*

$$\left|f\left(\frac{a+b}{2}\right)\right| \leqslant \frac{1}{b-a}\int_a^b |f(x)|\,dx + \frac{1}{2}\int_a^b |f'(x)|\,dx \ .$$

LEMMA 4. *Let* $\alpha(n)$ *and* $a(n)$ *be sequences of real and complex numbers respectively. Suppose both sequences are supported on* $[M,M']$, *where* $M' \leqslant M$. *Define*

$$F(t) = \sum_{M<n\leqslant M'} a(n)e(t\alpha(n)) \ .$$

Then for any $\delta > 0$,

$$\int_{-1/2\delta}^{1/2\delta} |F(t)|^2 dt \leqslant \left(\frac{\pi}{2}\right)^4 \frac{1}{\delta} \sum_n \sum_m a(n)\overline{a(m)}g(\delta^{-1}(\alpha(n)-\alpha(m))) \ .$$

Proof. Let

$$G(t) = \left(\frac{\pi}{2}\right)^2 \left(\frac{\sin \pi t\delta}{\pi t\delta}\right)^2 \ .$$

Then $G(t) \geqslant 1$ for $|t| \leqslant 1/2\delta$, so

$$\int_{-1/2\delta}^{1/2\delta} |F(t)|^2 dt \leqslant \int_{-\infty}^{\infty} |FG(t)|^2 dt \ ,$$

and by Plancherel's identity, this is

$$= \int_{-\infty}^{\infty} |\widehat{FG}(x)|^2 dx$$

$$= \left(\frac{\pi}{2}\right)^4 \delta^{-1} \sum_n \sum_m a(n)\overline{a(m)}g(\delta^{-1}(\alpha(n)-\alpha(m))) \ .$$

4. *Reformulation of Theorem 1 and initial stages of the proof.* Instead of working with $\zeta(s)$ directly, we use

Lemma 2 to go from Dirichlet polynomials to $\zeta(s)$. Accordingly, we shall prove

THEOREM 2. *For fixed positive numbers* N *and* σ *satisfying* $1/2 \leqslant \sigma \leqslant 1$, *define*

$$S(t) = S(t,N,\sigma) = \sum_n g(\log \frac{n}{N})n^{-\sigma} e(t \log n) . \tag{4.1}$$

Suppose $t_1,\ldots,t_R$ *are real numbers such that*

$$T < t_m \leqslant 2T, \quad |t_m - t_n| \geqslant 1 \quad (m \neq n) ,$$

and

$$|S(t_m)| \geqslant V .$$

Then

$$R \ll T^{14}V^{-196}N^{140-196\sigma}\mathcal{L}^{227} . \tag{4.2}$$

Note that $S(t)$ is a finite sum since $g(\log(n/N))$ is supported on $[Ne^{-2},Ne^2]$.

We deal first with some trivial cases of Theorem 2. The result is trivial if the right hand side of (4.2) is $\geqslant T$. Accordingly, we may assume that

$$V \geqslant T^{13/196}N^{5/7-\sigma} . \tag{4.3}$$

On the other hand, the exponent pair $(1/14,11/14)$ leads to the estimate $S(t) \ll T^{1/14}N^{5/7-\sigma}$, so we may assume that

$$V \ll T^{1/14}N^{5/7-\sigma} . \tag{4.4}$$

From the exponent pairs $(1/6,2/3)$ and $(0,1)$ we obtain $S(t) \ll T^{1/6}N^{1/2-\sigma}$ and $S(t) \ll N^{1-\sigma}$ respectively. Consequently (4.3) is satisfied only if

$$T^{13/56} \leqslant N \leqslant T^{59/126} < T^{1/2} . \tag{4.5}$$

We henceforth assume that (4.3), (4.4), and (4.5) are satisfied.

Let $\mathcal{S}_0 = \{t_1, \ldots, t_R\}$ be the set of points described in Theorem 2. For $\tau \in [T, 2T]$ define

$$\mathcal{S}(\tau) = \mathcal{S}_0 \cap [\tau - \tfrac{1}{2} G, \tau + \tfrac{1}{2} G] \quad .$$

Here G is a parameter to be chosen later; for now we assume only that

$$G \geq 1 \ . \tag{4.6}$$

If $t \in \mathcal{S}(\tau)$, then

$$V^2 \leq |S(t)|^2 \leq \int_{t-\frac12}^{t+\frac12} |S(a)|^2 da + \{\int_{t-\frac12}^{t+\frac12} |S'(a)|^2 da \int_{t-\frac12}^{t+\frac12} |S(a)|^2 da\}^{\frac12}$$

by Lemma 3 and the Cauchy-Schwarz inequality. Applying Cauchy-Schwarz again in the form $|ab| \leq \frac{1}{2}(|a|^2 + |b|^2)$, we get

$$V^2 \leq |S(t)|^2 \leq \frac{3}{2} \int_{t-\frac12}^{t+\frac12} |S(a)|^2 da + \frac{1}{2} \int_{t-\frac12}^{t+\frac12} |S'(a)|^2 da \quad .$$

Summing over all $t \in \mathcal{S}(\tau)$, we find that

$$V^2 |\mathcal{S}(\tau)| \leq \frac{3}{2} \int_{\tau-G}^{\tau+G} |S(t)|^2 dt + \frac{1}{2} \int_{\tau-G}^{\tau+G} |S'(t)|^2 dt \ .$$

Now we apply Lemma 4 with $\delta = 1/(2G)$ to both integrals. Upon writing $n + h$ for m, we obtain

$$|\mathcal{S}(\tau)| \leq GV^{-2} N^{-2\sigma} \mathcal{L}^2 \sum_n \sum_h a(n,h) e\left(\tau \log(\frac{n+h}{n})\right) \ , \tag{4.7}$$

where

$$a(n,h) = \frac{N^{2\sigma}}{\mathcal{L}^2} \left\{ \frac{g(\log(\frac{n+h}{N}))g(\log(\frac{n}{N}))g(2G \log(\frac{n+h}{n}))}{n^{\sigma}(n+h)^{\sigma}} \right\}$$

$$\cdot \{ c_1 + c_2 \log(n) \log(n+h) \} \cdot$$

Note that $a(h,h) \ll 1$ and $a(n,h) = 0$ if $\left| \log\left(\frac{n+h}{n}\right) \right| \geq 1/G$. Consequently, we may restrict h by the condition $|h| \leq H$, where $H \simeq N/G$. The terms with $h = 0$ contribute $\ll GV^{-2}N^{1-2\sigma}\mathcal{L}^2$ to the right hand side of (4.7). We choose G so that this contribution is $\leq 1/2$; i.e., we set

$$(4.8) \qquad\qquad G = c_3 V^2 N^{2\sigma-1} \mathcal{L}^{-2} \; ,$$

where c_3 is a sufficiently small positive constant. Note that (4.6) is satisfied if V satisfies (4.3). Consequently

$$|\mathcal{S}(\tau)| \leq 2 \, c_3 N^{-1} \sum_{1 \leq |h| \leq H} \sum_n a(n,h) e\left(\tau \, \log\left(\frac{n+h}{n}\right)\right)$$

$$= 2 \, c_3 N^{-1} \sum_{1 \leq |h| \leq H} S_1(\tau,h) \; ,$$

say. By the Cauchy-Schwarz inequality,

$$|\mathcal{S}(\tau)|^2 \leq 8 \, c_3^2 H N^{-2} \sum_{1 \leq |h| \leq H} |S_1(\tau,h)|^2$$

$$= c_4 \, G^{-1} N^{-1} \sum_{1 \leq |h| \leq H} |S_1(\tau,h)|^2 \; .$$

The previous argument is analogous to process A of the exponent pairs method. We now repeat this argument. Let $W \geq 1$ be a real number, and define

$$\mathcal{T}_0 = \mathcal{T}_0(W) = \{T + nG : 0 < n \leq T/G \text{ and } |\mathcal{S}(T+nG)| \geq W\} \; .$$

Let J be a real number to be chosen later; for now we assume only that

$$(4.9) \qquad\qquad\qquad J \geq G \; .$$

For $u \in [T,2T]$, define $\mathscr{T}(u) = \mathscr{T}_0 \cap [u-\frac{1}{2}J,\ u+\frac{1}{2}J]$. By Lemma 3,

$$|S_1(t,h)|^2 \leq \frac{1}{G}\int_{t-\frac{1}{2}G}^{t+\frac{1}{2}G} |S_1(\alpha,h)|^2 d\alpha + \int_{t-\frac{1}{2}G}^{t+\frac{1}{2}G} |S_1S_1'(\alpha,h)|\,d\alpha \quad .$$

Summing over points in $\mathscr{T}(u)$, we find that

$$(4.10)\qquad |\mathscr{T}(u)| \leq \frac{c_4}{GNW^2} \sum_{1\leq|h|\leq H} \int_{u-J}^{u+J} \left(\frac{1}{G}|S_1(t,h)|^2 + |S_1S_1'(t,h)|\right)dt.$$

By Lemma 4 with $\delta = 1/(2J)$

$$(4.11)\qquad \frac{1}{G}\int_{u-J}^{u+J} |S_1(t,h)|^2 dt \leq \frac{c_5J}{G} \sum_{|k|\leq K} \sum_n \beta_1(n,h,k)e(f(u,n,h,k)) \ ,$$

where

$$f(u,n,h,k) = u \log\left(\frac{(n+h)(n+k)}{n(n+h+k)}\right) \ ,$$

$$\beta_1(n,h,k) = a(n,h)a(n,k)g\left(2J \log\left(\frac{(n+h)(n+k)}{n(n+h+k)}\right)\right) \quad ,$$

and K satisfies

$$(4.12)\qquad \frac{HK}{N^2} \simeq \frac{1}{J} \quad .$$

The combination of the Cauchy-Schwarz inequality and Lemma 3 yields

$$(4.13)\qquad \int_{u-J}^{u+J} |S_1S_1'(t,h)|dt$$

$$\leq \left\{\int_{u-J}^{u+J} |S_1(t,h)|^2 dt\right\}^{\frac{1}{2}} \left\{\int_{u-J}^{u+J} |S_1'(t,h)|^2 dt\right\}^{\frac{1}{2}}$$

$$\leq c_6J \left\{ \sum_{|k|\leq K} \sum_n \beta_1(n,h,k)e(f(u,n,h,k))\right\}^{\frac{1}{2}} \times$$

$$\times \left\{ \sum_{|k| \leq K} \sum_{n} \beta_2(n,h,k)e(f(u,n,h,k)) \right\}^{\frac{1}{2}} ,$$

where

$$\beta_2(n,h,k) = \beta_1(n,h,k)\log(\frac{n+h}{n})\log(\frac{n+k+h}{n+k}) \ll G^{-2} .$$

Now we let $\beta_3 = G^2\beta_2$, and we use Cauchy-Schwarz in the form $|ab| \leq \frac{1}{2}(|a|^2 + |b|^2)$. An examination of the proof of Lemma 4 shows that the sums on the right hand side of (4.13) are positive, so

$$(4.14) \qquad \int_{u-J}^{u+J} |S_1S_1'(t,h)|dt \leq \frac{c_6J}{2G} \sum_{|k| \leq K} (\beta_1 + \beta_3)(n,h,k)e(f(u,n,h,k)).$$

The combination of (4.10), (4.11), and (4.14) yields

$$(4.15) \qquad |\mathcal{T}(u)| \leq \frac{J}{G^2NW^2} \sum_{1 \leq |h| \leq H} \sum_{|k| \leq K} \sum_{n} \beta(n,h,k)e(f(u,n,h,k)) ,$$

where

$$\beta(n,h,k) = \beta_1(n,h,k)\left(c_7 + c_8G^2\log(\frac{n+h}{n})\log(\frac{n+k+h}{n+k})\right) \ll 1 .$$

Now the terms with $k = 0$ contribute $\ll JHG^{-2}W^{-2}$ to the right hand side of (4.15). We choose J so that this contribution is $\leq 1/2$, i.e., we take

$$(4.16) \qquad J = c_9W^2G^2H^{-1} \simeq W^2V^6N^{6\sigma-4}\mathcal{L}^{-6} ,$$

where c_9 is any sufficiently small positive constant. Note that (4.9) is satisfied since we are assuming (4.3) and (4.5). We therefore obtain

$$(4.17) \qquad |\mathcal{T}(u)| \leq 2c_9H^{-1}N^{-1} \sum_{1 \leq |h| \leq H} \sum_{1 \leq |k| \leq K} \sum_{n} \beta(n,h,k)e(f(u,n,h,k)) .$$

Further analysis depends on the Poisson summation formula, and this forms the substance of the next section.

5. Application of the Poisson summation formula. We begin by splitting the sums over h and k in (4.17) into subintervals. Let $A = \log H/\log 2$, $B = \log K/\log 2$, $H = H2^{-a}$, and $K_b = K2^{-b}$. Then (4.17) may be written as

$$(5.1) \qquad |\mathscr{T}(u)| \leqslant 2c_9 H^{-1} N^{-1} \sum_{\substack{0 \leqslant a \leqslant A \\ 0 \leqslant b \leqslant B}} \{S_1(a,b,u) + \ldots + S_4(a,b,u)\} \quad ,$$

where

$$S_\nu(a,b,u) = \sum_{\frac{1}{2}H < \pm h < H_a} \; \sum_{\frac{1}{2}K_b < \pm k \leqslant K_b} \; \sum_h \beta(n,h,k)\, e(f(u,n,h,k))$$

and each S_ν $(1 \leqslant \nu \leqslant 4)$ corresponds to one choice of $\pm$ signs. We let S_1 correspond to the sum with both h and k positive. In the subsequent analysis we deal only with S_1, for the other sums may be treated in a very similar fashion.

Define

$$L = T\,H\,K\,N^{-3}$$

and

$$L_{a,b} = T\,H_a\,K_b\,N^{-3} = L2^{-a-b} \; .$$

There are absolute constants c_{10} and c_{11} such that if $\frac{1}{2}H_a < |h| \leqslant H_a$, $\frac{1}{2}K_b < |k| \leqslant |K_b|$, $T < u \leqslant 2T$, and $Ne^{-2} \leqslant x \leqslant Ne^2$, then

$$2c_{10}L_{a,b} \leqslant |f_x(u,x,h,k)| = \left|2u \int_0^h \int_0^k \frac{dv\,dw}{(x+h+k)^3}\right| \leqslant \frac{1}{2}c_{11}L_{a,b} \; .$$

LEMMA 5. Suppose $\frac{1}{2}H_a < h \leqslant H_a$, $\frac{1}{2}K_b < k \leqslant K_b$, and $T < u \leqslant 2T$. If $c_{11}L_{a,b} \geqslant 1$ then

$$(5.2) \qquad \sum_n \beta(n,h,k)\, e(f(u,n,h,k))$$

$$= \sum_{c_{10}L_{a,b} \leqslant \ell \leqslant c_{11}L_{a,b}} \int_{Ne^{-2}}^{Ne^{2}} \beta(x,h,k)e(f(u,x,h,k)+\ell x)dx$$

$$+ O(N^{-1}) \quad .$$

If $c_{11}L_{a,b} < 1$ *then*

(5.3) $$\sum_n \beta(n,h,k)e(f(u,n,h,k)) \ll L_{a,b}^{-1}$$

Proof. For simplicity of notation, we regard h,k and u as fixed, and we write $\beta(x)$ and $f(x)$ in place of $\beta(x,h,k)$ and $f(u,x,h,k)$ respectively. We also assume that $a = b = 0$.

By the Poisson summation formula,

$$\sum_n \beta(n)e(f(n)) = \sum_\ell \int_{Ne^{-2}}^{Ne^{2}} \beta(x)e(f(x)+\ell x)dx = \sum_\ell \Phi(\ell) \ ,$$

say. To complete the proof of (5.2), it suffices to estimate the contribution of those terms with $\ell < c_{10}L$ and $\ell > c_{11}L$. In either case,

$$|f'(x) + \ell| \gg |\ell| + L$$

for $Ne^{-2} \leqslant x \leqslant Ne^{2}$. Now let $\eta(x) = (f(x) + \ell x)^{-1}$, $\xi_0(x)=\beta(x)$, and $\xi_{\nu+1}(x) = (\xi_\nu(x)\eta(x)'$ for $\nu = 0$ or 1. Integrating by parts twice, we obtain

$$\Phi(\ell) = \left(\frac{-1}{2\pi i}\right)^2 \int_{Ne^{-2}}^{Ne^{2}} \xi_2(x)e(f(x) + \ell x)dx \ .$$

Here we are using the fact that β is twice differentiable. Moreover, $\beta^{(\nu)}(x) \ll N^{-\nu}$ for $0 \leqslant \nu \leqslant 2$, so

$$\xi_2 = \xi_0(\eta')^2 + \xi_0\eta\eta'' + 3\xi_0'\eta\eta' + \xi_0''\eta^2 \ll N^{-2}(|\ell|+L)^{-2} \quad .$$

Consequently, $\Phi(\ell) \ll N^{-1}(|\ell|+L)^{-2}$. Summing over ℓ completes the proof of (5.2).

If $c_{11}L < 1$, then the above argument shows that

$$\sum_n \beta(n)e(f(n)) = \Phi(0) + O(N^{-1}) .$$

If we integrate by parts once, we obtain

$$\Phi(0) = -\frac{1}{2\pi i}\int_{Ne^{-2}}^{Ne^{2}} \xi_1(x)e(f(x))dx \ll L^{-1} .$$

Since $L^{-1} \gg 1 \gg N^{-1}$, this proves (5.3).

The integrals $\Phi(\ell)$ with $c_{10}L < \ell \leqslant c_{11}L$ can be treated by the method of stationary phase. While there are several different versions of this in the literature, the simple version given in Titchmarsh's book suffices for our purposes.

LEMMA 6. *Let* h,k,u,β,f *and* Φ *be as in the proof of Lemma* 5. *Suppose that* $c_{11}L_{a,b} \geqslant 1$. *Define* $\psi = \psi(h,k,\ell,u)$ *by the relation*

$$-f'(\psi) = 2u\int_0^h\int_0^k \frac{dvdw}{(\psi+v+w)^3} = \ell . \tag{5.4}$$

Then

$$\Phi(\ell) = \frac{e(\frac{1}{8}+f(\psi)+\ell\psi)\beta(\psi)}{\sqrt{f''(\psi)}} + O(N^{2/5}L_{a,b}^{-3/5}) .$$

Proof. We again consider only the case $a = b = 0$ in order to simplify notation. Let

$$\beta_1(x) = \beta(x) - \beta(\psi) = \int_\psi^x \beta'(w)dw .$$

Then

$$\Phi(\ell) = \beta(\psi)\int_{Ne^{-2}}^{Ne^{2}} e(f(x)+\ell x)dx$$

$$+ \int_{Ne^{-2}}^{Ne^{2}} \beta_1(\psi)e(f(x)+\ell x)dx = \Phi_1 + \Phi_2 ,$$

say. By Lemma 4.6 of Titchmarsh [7],

$$(5.5)\qquad \Phi_1 = \frac{\beta(\psi)e(\frac{1}{8}+f(\psi)+\ell\psi)}{\sqrt{f''(\psi)}} + O(N^{2/5}L^{-3/5})$$

$$+ \; O\left(\frac{\beta(\psi)}{|f'(Ne^{-2})+\ell|} + \frac{\beta(\psi)}{|f'(Ne^{2})+\ell|}\right).$$

Now

$$\frac{\beta(\psi)}{|f'(Ne^{-2})+\ell|} = \frac{\int_{Ne^{-2}}^{\psi}\beta'(w)dw}{|\int_{Ne^{-2}}^{\psi}f''(w)dw|} \ll \frac{1}{L}$$

since $\beta'(w) \ll N^{-1}$ and $f''(w) \ll LN^{-1}$. Similarly

$$\frac{\beta(\psi)}{|f'(Ne^{2})+\ell|} \ll \frac{1}{L}\quad .$$

Since $L \geqslant 1$, the second "O" term in (5.5) may be absorbed into the first.

To estimate Φ_2, let δ be a parameter to be chosen later, and write

$$\Phi_2 = \int_{|x-\psi|\leqslant\delta} + \int_{|x-\psi|>\delta} = \Phi_3 + \Phi_4\ ,$$

say. Since $\beta_1(x) \ll |x-\psi|N^{-1}$, we see that

$$\Phi_3 \ll \delta^2 N^{-1}\quad .$$

For Φ_4, we observe that

$$\int_{\psi+\delta}^{Ne^2}\beta_1(x)e(f(x)+\ell x)dx$$

$$= \frac{1}{2\pi i}\int_{\psi+\delta}^{Ne^2}\beta_1(x)(f'(x)+\ell)^{-1}de(f(x)+\ell x)$$

$$\ll \int_{\psi+\delta}^{Ne^2}\left|\frac{d}{dx}\,\frac{\beta_1(x)}{f'(x)+\ell}\right|dx \ll \frac{N}{L\delta}$$

The same estimate holds for the integral on $[Ne^{-2}, \psi - \delta]$. Consequently

$$\Phi_4 \ll NL^{-1}\delta^{-1}$$

and

$$\Phi_2 \ll \delta^2 N^{-1} + NL^{-1}\delta^{-1} .$$

Upon taking $\delta = N^{2/3}L^{-1/3}$, we see that

$$\Phi_2 \ll N^{1/3}L^{-2/3} \ll N^{2/5}L^{-3/5} .$$

This, together with the estimate for Φ_1, completes the proof.

The application of the Halász method in the next section requires more explicit information about $f''(\psi)$. Since $h \leq H$ and $k \leq K \ll H^2$, we have $\psi + v + w = \psi + O(H^2)$ in the integral in (5.4). Therefore

$$\frac{2hk}{\psi^3}\{1 + O(H^2N^{-1})\} = \ell ,$$

i.e.

$$\psi = \left(\frac{2hk}{\ell}\right)^{1/3}\{1 + O(H^2N^{-1})\} .$$

It follows that

$$f''(\psi) = 6\int_0^h\int_0^k \frac{du\,dv}{(\psi + u + v)^4} \tag{5.6}$$

$$= \frac{3\ell^{4/3}}{(2uhk)^{1/3}}\{1 + O(H^2N^{-1})\} .$$

We must also express $\beta(\psi)$ in a different form; we do this by means of Fourier transforms. Let

$$\gamma(x) = \gamma(x,h,k) = \beta(Ne^x) ,$$

so that $\operatorname{supp}\gamma \subseteq [-2,2]$, γ has two continuous derivatives, and $\gamma^{(\nu)}(x) \ll 1$ for $0 \leq \nu \leq 2$. By the Fourier inversion formula,

$$(5.7) \qquad \beta(\psi) = \Upsilon(\log(\psi/N)) = \int_{-\infty}^{\infty} \hat{\gamma}(t)e(t \log (\psi/N))dt .$$

We also need an upper bound for $\hat{\gamma}(t)$. Note that

$$\hat{\gamma}(t) = \int_{-2}^{2} \gamma(x)e(-xt)dx \ll 1 .$$

Integrating by parts twice, we obtain

$$\hat{\gamma}(t) = \left(\frac{-1}{2\pi it}\right)^2 \int_{-2}^{2} \gamma(x)e(-tx)dt \ll |t|^{-2} ,$$

so

$$(5.8) \qquad \hat{\gamma}(t) \ll (1+|t|)^{-2} .$$

We rewrite (5.1) as

$$(5.9) \qquad |\mathcal{T}(u)| \leq 2c_9H^{-1}N^{-1} \Big\{ \sum_{(a,b)\in \mathcal{A}} + \sum_{(a,b)\in \mathcal{B}} \Big\}$$

$$\cdot (S_1(a,b,u) + \ldots + S_4(a,b,u)) ,$$

where $\mathcal{A}$ consists of those pairs (a,b) with $c_{11}L_{a,b} \geq 1$ and $\mathcal{B}$ consists of the remaining pairs. If $(a,b) \in \mathcal{B}$ then by (5.3) of Lemma 5,

$$S_\nu \ll H_aK_bL_{a,b}^{-1} \ll N^3T^{-1} ,$$

whence

$$H^{-1}N^{-1} \sum_{(a,b)\in \mathcal{B}} \{S_1 + \ldots + S_4\} \ll N^2H^{-1}T^{-1}\mathcal{L} \ll N^{2\sigma}V^2T^{-1} .$$

By (4.4) and (4.5), the above is

$$\ll N^{10/7}T^{-6/7} \ll T^{-1/7} .$$

Thus the contribution of terms with $(a,b) \in \mathcal{B}$ may be absorbed into the left hand side at the expense of increasing the constant on the right. In short,

$$(5.10) \qquad |\mathscr{T}(u)| \leqslant 3c_9 H^{-1}N^{-1} \sum_{(a,b)\in \mathfrak{A}} \{S_1(a,b,u) + \ldots + S_4(a,b,u)\} .$$

Let

$$(5.11) \qquad \Gamma_I(a,b,u) = \sum_{\frac{1}{2}H_a < h \leqslant H_a} \ \sum_{\frac{1}{2}K_b < k \leqslant K_b} (uhk)^{1/6} \cdot$$

$$\cdot \int_{-\infty}^{\infty} \hat{\gamma}(t) \sum_{c_{10}L_{a,b} < \ell \leqslant c_{11}L_{a,b}} \ell^{-2/3} e(f(\psi) + \ell\psi + t \log \frac{\psi}{N}) dt .$$

By (5.2) of Lemma 5, Lemma 6, (5.6), and (5.7),

$$S_1(a,b,u) = 2^{1/6}3^{1/2} e(1/8)\Gamma_1(a,b,u)$$

$$+ O(HKL^{2/5}N^{2/5} + H^3K\,L^{1/2}N^{-1/2}) .$$

Similar expressions may be obtained for S_2, S_3, and S_4. Consequently,

$$|\mathscr{T}(u)| \ll H^{-1}N^{-1} \sum_{(a,b)\in \mathfrak{A}} \{\Gamma_1(a,b,u) + \ldots + \Gamma_4(a,b,u)\}$$

$$+ O(K\,L^{2/5}N^{-3/5}\mathcal{L}^2 + H^2K\,L^{1/2}N^{-3/2}\mathcal{L}^2) .$$

By (4.3) through (4.5), the error terms may be absorbed into the left hand side, so

$$(5.12) \qquad |\mathscr{T}(u)| \ll H^{-1}N^{-1} \sum_{(a,b)\in \mathfrak{A}} \{|\Gamma_1(a,b,u)| + \ldots + |\Gamma_4(a,b,u)|\} .$$

At this point it is instructive to note that, by the trivial estimate for Γ_ν, we have

$$(5.13) \qquad |\mathscr{T}(u)| \ll H^{-1}N^{-1}H\,K\,L(THKL^{-4})^{1/6} \simeq K\,L^{1/2}N^{-1/2}$$

$$\simeq T^{1/2}N^{5-7\sigma}\mathcal{L}^7V^{-7}W^{-3} .$$

When $W = 1$ and $V \gg T^{1/14}N^{5/7-\sigma}\mathcal{L}$, we get $|\mathcal{T}(u)| = 0$. Thus (5.13) essentially contains (1.3). The advantage of (5.13) is that it can be combined with Halász's large values method, and this is the subject of the next section.

6. Application of Halász's Method. Let $T_0 \in [T, 2T]$ and let Q be a parameter to be chosen later. Define

$$\mathcal{T}_2 = \{u = T_0 + nJ : 0 \leqslant n \leqslant Q/J \text{ and } \mathcal{T}(u) \neq \emptyset\} \quad .$$

We begin by observing that by (5.12) and the Cauchy-Schwarz inequality,

$$(6.1) \qquad \sum_{u \in \mathcal{T}_2} |\mathcal{T}(u)| \leqslant \sum_{u \in \mathcal{T}_2} |\mathcal{T}(u)|^2$$

$$\ll H^{-2}N^2\mathcal{L}^2 \sum_{(a,b) \in \mathcal{Q}} \sum_{u \in \mathcal{T}_2} \cdot$$

$$\cdot \{|\Gamma_1(a,b,u)|^2 + \ldots + |\Gamma_4(a,b,u)|^2\} .$$

We concentrate on estimating

$$\sum_{u \in \mathcal{T}_2} |\Gamma_\nu(a,b,u)|^2 \quad .$$

For simplicity of notation, we assume that $\nu = 1$ and $a = b = 0$.

Let

$$(6.2) \quad \varphi(u,h,k,\ell,t) = u \log \frac{(\psi + h)(\psi + k)}{\psi(\psi + h + k)} + \ell\psi + t \log(\psi/N) \quad ,$$

where ψ is as defined in (5.4), and let

$$F(u,h,\ell,t) = \sum_{\frac{1}{2}K < k \leqslant K} k^{1/6} \hat{\gamma}_{h,k}(t) e(\varphi(u,h,k,\ell,t)) \; .$$

(Note that $\hat{\gamma}$ depends on h,k, and t but not on ℓ and u.) Then

$$\Gamma_1(0,0,u) = u^{1/6} \sum_{\frac{1}{2}H<h\leqslant H} \sum_{c_{10}L<\ell\leqslant c_{11}L} h^{1/6}\ell^{-2/3} \int_{-\infty}^{\infty} F(u,h,\ell,t)\,dt \quad .$$

Two applications of Cauchy-Schwarz give us

$$|\Gamma_1(0,0,u)|^2 \ll T^{1/3}H^{4/3}L^{-1/3} \sum_h \sum_\ell \left|\int_{-\infty}^{\infty} F\,dt\right|^2$$

and

$$\left|\int_{-\infty}^{\infty} F\,dt\right|^2 \leqslant \int_{-\infty}^{\infty} (1+|t|)^{-2}dt \int_{-\infty}^{\infty} (1+|t|)^2|F|^2dt \quad .$$

Consequently

$$(6.3) \quad \sum_{u\in \mathcal{T}_2} |\Gamma_1(0,0,u)|^2 \ll T^{1/3}H^{4/3}L^{-1/3} \sum_{\frac{1}{2}H<h\leqslant H} \sum_{c_{10}L<\ell\leqslant c_{11}L}$$

$$\cdot \int_{-\infty}^{\infty} (1+|t|)^2 \sum_{u\in \mathcal{T}_2} |F(u,h,\ell,t)|^2dt \ .$$

Now we fix h,ℓ, and t and bound, via Halász's method, the sum

$$(6.4) \quad \sum_{u\in \mathcal{T}_2} |F(u,h,\ell,t)|^2 = \sum_{u\in \mathcal{T}_2} \Big| \sum_{\frac{1}{2}K<k\leqslant K} \hat{\gamma}_k k^{1/6} e(\varphi(u,k)) \Big|^2 \quad .$$

Here we have written $\hat{\gamma}_k$ for $\hat{\gamma}_{h,k}(t)$ and $\varphi(u,k)$ for $\varphi(u,h,k,\ell,t)$.

Instead of appealing to the version of Halász's inequality in Montgomery's book, we appeal to a well known duality principle ([5], Lemma 2.) It implies that the following two assertions about the constant D are equivalent:

(6.5)i <u>For any</u> a(k),

$$\sum_{u\in\mathcal{T}_2} \Big| \sum_{\frac12 K<k\leqslant K} a(k)e(\varphi(u,k))\Big|^2 \leqslant D \sum_{\frac12 K<k\leqslant K} |a(k)|^2 .$$

(6.5)ii <u>For any</u> $b(u)$,

$$\sum_{\frac12 K<k\leqslant K} \Big| \sum_{u\in\mathcal{T}_2} b(u)e(\varphi(u,k))\Big|^2 \leqslant D \sum_{u\in\mathcal{T}_2} |b(u)|^2 .$$

In the next lemma we prove a bound for these bilinear forms by considering the second one.

LEMMA 7. <u>Let</u> (κ,λ) <u>be an exponent pair</u>. <u>Then there is a constant</u> c_{12} <u>such that if</u> $|t| \leqslant c_{12}T/J$ <u>then</u> (6.5)i <u>and</u> (6.5)ii <u>are true for some</u> D <u>satisfying</u>

$$D \ll K\mathcal{L} + Q^{\kappa} J^{-\kappa} K^{\lambda-\kappa}|\mathcal{T}_2| .$$

<u>Proof</u>. The left hand side of (6.5)ii is

$$= \sum_{u_1\in\mathcal{T}_2} \sum_{u_2\in\mathcal{T}_2} b(u_1)\overline{b(u_2)} \sum_{\frac12 K<k\leqslant K} e(\varphi(u_1,k)-\varphi(u_2,k))$$

$$\leqslant \sum_{u_1\in\mathcal{T}_2} \sum_{u_2\in\mathcal{T}_2} \left(\frac12|b(u_1)|^2 + \frac12|b(u_2)|^2\right)\cdot$$

$$\cdot \Big| \sum_{\frac12 K<k\leqslant K} e(\varphi(u_1 k) - \varphi(u_2,k))\Big|$$

$$= \sum_{u_1\in\mathcal{T}_2} |b(u_1)|^2 \sum_{u_2\in\mathcal{T}_2} \Big| \sum_{\frac12 K<k\leqslant K} e(\varphi(u_1,k) - \varphi(u_2,k))\Big| .$$

To estimate the inner sum, we need information about

$$\frac{\partial}{\partial k}(\varphi(u_1,k) - \varphi(u_2,k)) = \int_{u_1}^{u_2} \varphi_{uk}(u,k)du .$$

Now

$$\varphi_k = \frac{uh}{\psi(\psi+h+k)} + \frac{t}{\psi}\psi_k ,$$

$$\frac{uh}{\psi(\psi+h+k)} \simeq \frac{TH}{N^2} \simeq \frac{T}{JK} ,$$

and

$$\frac{t}{\psi}\psi_k \simeq \frac{|t|}{K} .$$

So there is a constant c_{12} such that if $|t| \leqslant c_{12}T/J$, then $\varphi_k \simeq T/JK$. Similarly, if $|t| \leqslant c_{12}T/J$ then $\varphi_{uk} \simeq 1/JK$ and

$$\frac{\partial}{\partial k}(\varphi(u_1,k) - \varphi(u_2,k)) \simeq \frac{|u_1 - u_2|}{JK} .$$

If

$$\left|\frac{\partial}{\partial k}(\varphi(u_1,k) - \varphi(u_2,k))\right| \leqslant 1/2 \tag{6.6}$$

for $\frac{1}{2}K < k \leqslant K$ then

$$\sum_{\frac{1}{2}K<k\leqslant K} e(\varphi(u_1,k) - \varphi(u_2,k)) \ll \min(K, JK|u_1 - u_2|^{-1})$$

by Lemma 4.8 of Titchmarsh [7] and the trivial estimate. If (6.6) fails to hold then

$$\sum_{\frac{1}{2}K<k\leqslant K} e(\varphi(u_1,k) - \varphi(u_2,k)) \ll J^{-\kappa}K^{\lambda-\kappa}|u_1 - u_2|^{\kappa}$$

by the theory of exponent pairs. Therefore we may take

$$D \ll K + \sum_{u_1\in\mathscr{T}_2}\ \sum_{\substack{u_2\in\mathscr{T}_2\\ u_1\neq u_2}} \{JK|u_1 - u_2|^{-1} + J^{-\kappa}K^{\lambda-\kappa}|u_1 - u_2|^{\kappa}\} .$$

Since the points of $\mathscr{T}_2$ are separated by at least J,

$$\sum_{\substack{u_2\in\mathscr{T}_2\\ u_2\neq u_1}} |u_1 - u_2|^{-1} \ll J^{-1} \sum_{1\leqslant r\leqslant Q/J} r^{-1} \ll \mathscr{L}J^{-1} .$$

Furthermore, since $|u_1 - u_2| \leqslant Q$, we have

$$\sum_{u_2 \in \mathcal{T}_2} |u_1 - u_2|^{\kappa} \ll Q^{\kappa} |\mathcal{T}_2| \quad .$$

This completes the proof.

If $|t| \leqslant c_{12}T/J$, then Lemma 7 and (6.5)i with $a(k) = \hat{\gamma}_k k^{1/6}$ give us

$$\sum_{u \in \mathcal{T}_2} |F(u,h,k,\ell,t)|^2 \ll (1+|t|)^{-4} K^{4/3} \{K\mathcal{L} + Q^{\kappa} J^{-\kappa} K^{\lambda-\kappa} |\mathcal{T}_2|\} \; .$$

If $|t| \geqslant c_{12}T/J$, then we have the trivial bound

$$\sum_{u \in \mathcal{T}_2} |F(u,h,k,\ell,t)|^2 \ll (1+|t|)^{-4} K^{7/3} |\mathcal{T}_2| \quad .$$

These estimates together with (6.3) yield

$$(6.7) \qquad \sum_{u \in \mathcal{T}_2} |\Gamma_1(0,0,u)|^2 \ll T^{1/3} H^{7/3} K^{7/3} L^{2/3}$$
$$\cdot \{\mathcal{L} + Q^{\kappa} J^{-\kappa} K^{\lambda-\kappa-1} |\mathcal{T}_2| + JT^{-1} |\mathcal{T}_2|\} \quad .$$

In much the same fashion, we may prove that

$$\sum_{u \in \mathcal{T}_2} |\Gamma_{\nu}(a,b,u)|^2 \ll T^{1/3} H_a^{7/3} K_b^{7/3} L_{a,b}^{2/3}$$
$$\cdot \{\mathcal{L} + Q^{\kappa} J^{-\kappa} K_b^{\lambda-\kappa-1} |\mathcal{T}_2| + JT^{-1} |\mathcal{T}_2|\} \quad .$$

We may now sum over a and b and use (6.1) to get an upper bound for $\sum_{u \in \mathcal{T}_2} |\mathcal{T}(u)|$. To simplify the statement of this bound, let

$$M = \frac{1}{HN}\left(\frac{THK}{L}\right)^{1/3} H^2 K^2 L^2 \simeq \frac{TN^{10-14\sigma} \mathcal{L}^{14}}{V^{14} W^6} \quad .$$

(Note that the trivial bound for $|\mathcal{T}(u)|$ in (5.13) is $\simeq M^{1/2}$.) Then

$$\sum_{u \in \mathcal{T}_2} |\mathcal{T}(u)| \ll M\mathcal{L}^3 + M\mathcal{L}^2 Q^{\kappa} J^{-\kappa} K^{\lambda-\kappa-1} |\mathcal{T}_2| + M\mathcal{L}^2 J T^{-1} |\mathcal{T}_2| . \tag{6.8}$$

Now

$$M J \mathcal{L}^2 T^{-1} \ll N^{6-8\sigma} V^{-8} W^{-2} \mathcal{L}^{11} \ll \mathcal{L}^{-1}$$

by (4.3) and (4.5), so the third term on the right hand side of (6.8) may be absorbed into the left hand side. We choose Q so that the second term may also be absorbed. Consequently, we take

$$Q = c_{14} J K^{1+1/\kappa-\lambda/\kappa} M^{-1/\kappa} \mathcal{L}^{-2/\kappa} ,$$

where c_{14} is a sufficiently small positive constant. With this choice of Q, we have

$$\sum_{u \in \mathcal{T}_2} |\mathcal{T}(u)| \ll M\mathcal{L}^3 . \tag{6.9}$$

Now we divide the interval $[T,2T]$ into $\leq T/Q + 1$ intervals of length $\leq Q$. We apply (6.9) to each one to get

$$|\mathcal{T}_0(W)| \ll M\mathcal{L}^3(1 + T/Q) \ll M\mathcal{L}^3 + M^{1/\kappa}\mathcal{L}^{3+2/\kappa} T M J^{-1} K^{\lambda/\kappa-1/\kappa-1} . \tag{6.10}$$

By taking $(\kappa,\lambda) = (1/14, 11/14)$, we get

$$|\mathcal{T}_0(W)| \ll M\mathcal{L}^3 + M^{14}\mathcal{L}^{31} T^2 V^{-4} N^{-2-4\sigma} \mathcal{L}^4 . \tag{6.11}$$

In deriving the estimate (6.11), we estimated the sum over ℓ trivially and the sum over k by the method of exponent pairs. Consequently (6.11) is weak when K is small. However, if K is small then L is relatively large. This means that (6.11) may be complemented by an estimate in which the sum over k is estimated trivially and the sum over ℓ is estimated non-trivially. An argument similar to that used in deriving (6.3) gives us

$$|\Gamma_1(0,0,u)|^2 \ll T^{1/3}H^{4/3}K^{4/3} \sum_{\frac{1}{2}H<h\leq H} \sum_{\frac{1}{2}K<k\leq K}$$

$$\cdot \int_{-\infty}^{\infty} |\hat{\gamma}(t)| \sum_{u\in \mathcal{T}_2} \Big| \sum_{c_{10}L<\ell\leq c_{11}L} \ell^{-2/3} e(\varphi(u,\ell,t))\Big|^2 dt .$$

By applying Halász's method in much the same way as before, we get

$$|\mathcal{T}_0(W)| \ll M\mathcal{L}^3 + M^{1/\kappa}\mathcal{L}^{3+2/\kappa} T M J^{-1} L^{\lambda/\kappa-1/\kappa-1}$$

for any exponent pair (κ,λ). We again take $(\kappa,\lambda) = (1/14, 11/14)$ to get

(6.12) $$|\mathcal{T}_0(W)| \ll M\mathcal{L}^3 + M^{14}\mathcal{L}^{31}T^{-2}V^4N^{2+4\sigma}\mathcal{L}^{-4} .$$

From (6.11) and (6.12) together, we get

(6.13) $$\mathcal{T}_0(W) \ll M\mathcal{L}^3 + M^{14}\mathcal{L}^{31} \ll \frac{TN^{10-14\sigma}\mathcal{L}^{17}}{V^{14}W^6} + \frac{T^{14}N^{140-196\sigma}\mathcal{L}^{227}}{V^{196}W^{84}} .$$

Note that in the most important case, namely $W = 1$, the second term dominates because of (4.4).

7. Completion of proofs. To prove Theorem 2, we use (6.13) and the observation that

$$|\mathcal{S}_0| \leq \sum_{k=0}^{\infty} |\mathcal{T}_0(2^k)| 2^{k+1} \ll T^{14}N^{140-196\sigma}V^{-196}\mathcal{L}^{227} .$$

Now we prove Theorem 1. By Lemma 2,

$$V \ll |\zeta(5/7+it_m)| \ll \sum_{N \in \mathcal{N}} |S(t_m/2\pi, N, 5/7)| + O(1) \quad .$$

We may assume that $V \geq T^{13/196}$, so that the $O(1)$ term may be absorbed into the left hand side. By Hölder's inequality,

$$RV^{196} \ll \sum_{m=1}^{R} |\zeta(5/7+it_m)|^{196}$$

$$\ll \sum_{m=1}^{R} \Big| \sum_{N \in \mathcal{N}} S(t_m/2\pi, N, 5/7)\Big|^{196}$$

$$\ll \mathcal{L}^{195} \sum_{N \in \mathcal{N}} \sum_{m=1}^{R} |S(t_m/2\pi, N, 5/7)|^{196} \quad .$$

It follows that there is some N for which

$$RV^{196}\, \mathcal{L}^{-196} \ll \sum_{m=1}^{R} |S(t_m/2\pi, N, 5/7)|^{196} \quad .$$

We apply Theorem 2 to get

$$RV^{196}\mathcal{L}^{-196} \ll \sum_{1 \leq k \ll \mathcal{L}} 2^{196k} |\{t_m : 2^{k-1} \leq |S(t_m/2\pi, N, 5/7)| < 2^k\}|$$

$$\ll \sum_{1 \leq k \ll \mathcal{L}} T^{14}\, \mathcal{L}^{227} \ll T^{14}\, \mathcal{L}^{228} \quad ,$$

and this proves the theorem.

8. Some concluding remarks. There is one aspect of Heath-Brown's twelfth-power estimate that we have not been able to exploit here. There, the function analogous to φ (defined in (6.2)) is

$$\Phi(u,h,\ell) = u \log \frac{\Psi + h}{\Psi} + \Psi,$$

where $\Psi = \Psi(u,h,\ell)$ is defined by

$$\frac{u}{\Psi} - \frac{u}{\Psi + h} = \ell \quad .$$

Note that $\Phi(u,h,\ell) = \Phi(u,h\ell,1)$, so the variables h and ℓ may effectively be combined into a single variable. This fact may be used to an advantage in Halász's method. If the variables k and ℓ in (6.3) could be similarly combined, then we could prove

$$\sum_{u \in \mathcal{T}_2} |\Gamma_1(0,0,u)|^2 \ll_\epsilon M^{1+\epsilon} + M^{1+\epsilon} Q^{\kappa} J^{-\kappa} (KL)^{\lambda-\kappa-1} |\mathcal{T}_2|$$

for any $\epsilon > 0$. This would improve Theorem 1 to

$$R \ll_\epsilon T^{2+\epsilon} V^{-28} \quad .$$

In "Exponent pairs and the zeta-function of Riemann" (Studia Scien. Hungarica 15 (1980), 157-181), A. Ivić proves

$$\int_0^T |\zeta(5/7 + it)|^{12}\, dt \ll T^{1+\epsilon} \quad .$$

In the notation of Theorem 1, this implies $R \ll T^{1+\epsilon} V^{-12}$. However, estimate (1.3) does not follow from Ivić's result.

NOTES

1 Supported in part by NSF Grant No. MCS-8002153.

Address:
Department of Mathematics
Michigan Technological University
Houghton, Michigan 49931

REFERENCES

[1] Atkinson, F.V., "The mean value of the Riemann zeta function", Acta Math. 81 (1949) 353 - 376.

[2] Heath-Brown, D.R., "The twelfth power moment of the Riemann zeta-function", Quart. J. Math. Oxford (2) 29 (1978) 443 - 462.

[3] Kolesnik, G., "On the order of $\zeta(\frac{1}{2} + it)$ and $\Delta(R)$", Pac. J. Math. 98 (1982), 107 - 122.

[4] Montgomery, H.L., Topics in multiplicative number theory, Springer, Berlin, 1971.

[5] ------, "The analytic principle of the large sieve", Bull. Amer. Math. Soc. 84 (1978) 547 - 567.

[6] Phillips, E., "The zeta-function of Riemann: further developments of van der Corput's method", Quart. J. Math. Oxford 4 (1933) 209 - 225.

[7] Titchmarsh, E.C., The theory of the Riemann zeta-function, Oxford, 1951.

On the Zeros (à la Selberg) of Dirichlet Series Attached to Certain Cusp Forms

James Lee Hafner

1. <u>Introduction</u>. Let $F(z)$ be a holomorphic cusp form of even integral weight k on the full modular group $\Gamma(1)$ with constant multiplier system. That is,

$$F(Mz) = (cz+d)^k F(z), \qquad M = \begin{pmatrix} * & * \\ c & d \end{pmatrix} \in \Gamma(1) .$$

Write $F(z)$ in a Fourier series at the cusp ∞ as

$$F(z) = \sum_{\ell=1}^{\infty} f(\ell) e(\ell z), \qquad e(z) = e^{2\pi iz} .$$

We assume $F(z)$ is an eigenfunction of all the Hecke operators and $f(1) = 1$. Consequenctly, $f(\ell)$ is a real valued multiplicative function satisfying

$$\text{(a)} \quad f(p^\lambda) = f(p) f(p^{\lambda-1}) - p^{k-1} f(p^{\lambda-2}), \; p \text{ prime}, \; \lambda \geqslant 1;$$

$$\text{(1.1)} \quad \text{(b)} \quad |f(p)| \leqslant 2\, p^{(k-1)/2}, \; p \text{ prime};$$

$$\text{(c)} \quad \sum_{\ell \leqslant x} |f(\ell)|^2 \ll x^k, \qquad x \geqslant 0 .$$

(We use the convention that $f(x) = 0$ if x is not a positive integer.)

The identity in (a) was conjectured by Ramanujan and proved by Mordell [9] for the special case

$$F(z) = \Delta(z) = e(z) \prod_{n=1}^{\infty} (1 - e(nz))^{24} = \sum_{\ell=1}^{\infty} \tau(\ell)\, e(\ell z) .$$

The general case was proved by Hecke [7]. The inequality (b), called the Ramanujan-Petersson inequality, was proved by Deligne [1]. The estimate in (c) was first proved by Hardy

[5] and later refined to an asymptotic formula by Rankin [10].

We form the Dirichlet series

$$L_f(s) = \sum_{\ell=1}^{\infty} f(\ell)\ell^{-s} \quad ,$$

and note that (1.1) implies

$$(1.2) \qquad L_f(s) = \prod_p (1 - f(p)p^{-s} + p^{k-1-2s})^{-1} \quad ,$$

with the product and the series converging absolutely for $\sigma > (k+1)/2$. Hecke [6] showed that these Dirichlet series satisfy the functional equation :

$$(1.3) \qquad \xi_f(s) = (2\pi)^{-s} \Gamma(s) L_f(s) = \xi_f(k-s)(-1)^{k/2} .$$

Also, $\xi_f(s)$ is an entire function of order 1 and infinite type. As with the Riemann zeta function, we study the zeros of $\xi_f(s)$, and so of $L_f(s)$.

The Euler product (1.2) and the functional equation (1.3) together with a result of Rankin [10] imply that all the zeros of $\xi_f(s)$ lie in the strip $(k-1)/2 < \sigma < (k+1)/2$. Put

$$N(T) = \# \{ \rho = \beta + i\gamma : 0 \leqslant \gamma \leqslant T ,$$
$$(k-1)/2 \leqslant \beta \leqslant (k+1)/2, \ L_f(\rho) = 0\},$$

and

$$N_0(T) = \# \{\rho = k/2 + i\gamma : 0 \leqslant \gamma \leqslant T, \ L_f(\rho) = 0\} .$$

Lekkerkerker [8] has shown that

$$N_0(T) \sim c\, T \log T \ , \quad N_0(T) > AT \ ; \quad T \to \infty$$

for some positive constants c and A . Of course the Riemann hypothesis for these functions would be

$$N(T) = N_0(T) , \qquad T > 0 .$$

In this paper we prove the following.

Theorem. For a cusp form F as above, there are constants A and T_0 depending on F such that

$$N_0(T) > AT \log T , \qquad T > T_0 .$$

This theorem is the analogue of Selberg's result [11] for the Riemann zeta function. It extends the class of Dirichlet series for which it is known that a positive proportion of their zeros lie on their critical line.

The method of proof uses three main ingredients. First, we modify slightly Selberg's idea to introduce a mollifier $\phi(s)$ which approximates $L_f^{-\frac{1}{2}}$. Secondly, we require an approximate functional equation for $\xi_f(s)$ which is provided by Good [2]. Finally, we need to extend Good's techniques [3] for computing the mean-square of $L_f(k/2 + it)$.

There are some extra difficulties which we encounter which make this theorem quite difficult. First, the coefficients $f(\ell)$ are not completely multiplicative. This makes certain arithmetical sums more difficult to analyze. See Section 5. Secondly, with the mollifier, we need to study the behavior of

$$\sum_{\ell=1}^{\infty} f(\ell)\, f\!\left(\frac{\ell b+n}{a}\right) (\ell b + n/2)^{-s}, \quad (a,b) = 1,\ n \geq 0 ,$$

in a region of non-absolute convergence. In particular, this behavior must be uniform in a,b and n. This is obtained by appealing to the spectral theory of the Laplacian on the space $L^2(\Gamma_0(a,b)\backslash\mathcal{H})$, where $\Gamma_0(a,b)$ is a certain congruence group defined in (4.5). In this context, any estimate we use involving the spectrum must be uniform in a and b . This is the main difficulty. However, we can appeal to recent work of the author [4] for the necessary estimates. See Lemma 4.1.

We shall proceed as directly as possible through the proof. Though the details are somewhat complicated, we hope the ideas are clear and the method is instructive. An alternate proof using the Selberg-Titchmarsh techniques will appear elsewhere.

2. Notation and Main Lemma. To estimate $N_0(T)$ we want to study the behavior of $\xi_f(k/2+it)$ on $0 < t < T$. We first introduce a mollifier to control its oscillations, without introducing any new zeros (of odd order). The mollifier is chosen to approximate $L_f^{-\frac{1}{2}}$.

Define a multiplicative function

$$a_\nu = \frac{\mu(\nu)\ f(\nu)}{d(\nu)}$$

where μ and d are the usual Möbius and divisor functions. In particular, note that a_ν is non-zero only on square-free numbers and for a prime p,

$$(2.1) \qquad a_p = -f(p)/2 \quad .$$

Now for $\xi \geqslant 3$, let

$$(2.2) \quad \beta_\nu = a_\nu(\log^+\xi/\nu)/\log\xi \ , \quad \log^+x = \max(0,\log x) \ ,$$

and define

$$\phi(s) = \phi_\xi(s) = \sum_{\nu=1}^{\infty} \beta_\nu \nu^{-s}$$

Because $\beta_\nu = 0$ if $\nu \geqslant \xi$, $\phi(s)$ is entire. The Euler product (1.2) and (2.1) show that ϕ does indeed approximate $L_f^{-\frac{1}{2}}$. Also Deligne's estimate (1.1b) implies

$$(2.3)\qquad |\beta_\nu| \leqslant |a_\nu| \leqslant \nu^{(k-1)/2} .$$

Next put

$$G(t) = \gamma(t)\, L_f(k/2+it)\, i^{k/2} |\phi(k/2 + it)|^2$$

with

$$\gamma(t) = (2\pi)^{-it}\, \Gamma(k/2+it)/|\Gamma(k/2+it)| .$$

Equation (1.3) and the fact that $f(\ell)$ is real imply that $G(t)$ is real-valued and that if $G(t)$ has a change of sign at $t = \gamma$ then $L_f(k/2+i\gamma) = 0$.

Our theorem is then a consequence of the following.

Lemma. If T is sufficiently large, $\xi = T^{1/70}$, $(\log \xi)^{-1} \leqslant h \leqslant (\log \xi)^{-9/11}$, then

$$\int_0^T |G(t)|^2 dt \ll \frac{T \log T}{\log \xi} ,$$

and

$$\int_0^T \Big| \int_t^{t+h} G(u)\,du \Big|^2 dt \ll \frac{h^{3/2} T}{\sqrt{\log \xi}} .$$

The deduction of the theorem from these estimates follows precisely as in Selberg's theorem [11] or as in Titchmarsh's version [12, ch. 10]. We give the key ideas. Let E be the subset of $(0,T)$ where

$$\left| \int_t^{t+h} G(u)\,du \right| < \int_t^{t+h} |G(u)|\,du .$$

Note that for t in E, $G(u)$ must have a change of sign in $(t,t+h)$ so that $L_f(k/2+iu)$ must vanish in this interval. With the estimates in the lemma and

$$h = A(\log \xi)^{-1}, \quad A \text{ a large constant,}$$

we deduce $m(E) \gg T$. From this we conclude

$$N_0(T) \gg m(E)/h \gg T \log T ,$$

by our choce of $\xi = T^{1/70}$.

Our task is then to prove the lemma. This in turn will be deduced from our Main Lemma given below. First, we need some notation.

Let Ψ be a real-valued function defined by

$$\Psi(t) = \int_{-\infty}^{t} \{\psi(\tau - 5/4) - \psi(\tau - 7/4)\}\, d\tau$$

$$= \int_{5/4}^{7/4} \psi(t-\tau)\, d\tau$$

where ψ is a non-negative, multiply differentiable function satisfying

$$\int_{-\infty}^{\infty} \psi(\tau)\, d\tau = 1 , \quad \psi(\tau) = 0 \text{ for } |\tau| \geqslant 3/4 .$$

Then

$$(2.4) \qquad \Psi(t) = \begin{cases} 0 & t \leqslant 1/2 \text{ or } t \geqslant 5/2 \\ 1 & 1 \leqslant t \leqslant 2 . \end{cases}$$

Furthermore, for $t \geqslant 0$, $0 \leqslant u$, $v \leqslant h \leqslant 1$, let

$$(2.5) \qquad E(t;u,v) = G(t+u)\, \overline{G(t+v)} .$$

<u>Main Lemma</u>. Under the assumptions of the lemma above,

$$\text{(A)} \quad I(T) = \int_{-\infty}^{\infty} \Psi(t/T)\, E(t;0,0)dt \ll \frac{T \log T}{\log \xi}$$

$$\text{(B)} \quad J(T,h) = \int_0^h \int_0^h \left\{ \int_{-\infty}^{\infty} \Psi(t/T)\, E(t;u,v)dt \right\} dudv \ll \frac{h^{3/2}T}{\sqrt{\log \xi}}$$

Clearly (2.4) together with (A) and (B) gives a proof of the first lemma and so the theorem.

3. First Expression for $I(T;u,v)$. We need to compute

$$I(T;u,v) = \int_{-\infty}^{\infty} \Psi(t/T) \quad E(t;u,v)dt$$

where $0 \leqslant u,v \leqslant h \leqslant 1$ and $E(t;u,v)$ is defined in (2.5). We proceed in the following way. First we give an expression for $G(t)$ which is essentially the approximate functional equation of Good [2] for $L_f(k/2+it)$ after the mollifier ϕ is incorporated. We remark that G is given with the factor $\gamma(t)$ so that the approximate functional equation is more symmetric. Then we proceed in the calculation directly using the techniques developed by Good [2,3].

Throughout the following calculations we shall use the notational symmetry between u and v, and be free to exchange their roles. This is legitimate because in the applications either $u = v = 0$ or u and v are independently integrated over the interval $(0,h)$. Also, all our estimates will be uniform in $0 \leqslant u, v \leqslant 1$. These conditions will be assumed without further remark.

To develop the modified approximate functional equation, we need some auxiliary functions. Let φ be a real-valued, multiply differentiable function defined on the real line such that

$$(3.1) \qquad \varphi(\tau) = \begin{cases} 0 & |\tau| \geqslant 3/2 \\ 1 & |\tau| \leqslant 2/3 \end{cases}$$

and let

$$\varphi^*(\tau) = 1 - \varphi(1/\tau) .$$

Note that φ^* satisfies (3.1) also. We shall have occasion later to introduce other functions of this general type and we establish the convention that if ρ has compact support then

$$(3.2) \qquad \rho^*(\tau) = \rho(0) - \rho(1/\tau) .$$

Then as in Good [2], with a slight change of notation,

$$G(t) = \Big(A(t,\varphi) + B(t,\varphi)\Big) i^{k/2} ,$$

where

$$A(t,\varphi) = \gamma(t) \sum_{\ell,\mu,\nu} \frac{f(\ell)\beta_\mu \beta_\nu}{(\ell\mu\nu)^{k/2}} \left(\frac{\ell\mu}{\nu}\right)^{-it} \varphi\left(\frac{2\pi\,\ell\mu}{\nu t}\right)$$

$$+ \gamma(-t) \sum_{\ell,\mu,\nu} \frac{f(\ell)\beta_\mu \beta_\nu}{(\ell\mu\nu)^{k/2}} \left(\frac{\ell\mu}{\nu}\right)^{it} \varphi^*\left(\frac{2\pi\,\ell\mu}{\nu t}\right)$$

and

$$(3.2) \qquad B(t,\varphi) \ll (|t|+1)^{-\frac{1}{2}}\, \xi \Big(\log \xi(|t|+1)\Big)^2 .$$

The proof of this is not fundamentally different from that of Good's approximate functional equation in [2]. It is only necessary to consider the expression

$$\phi(s)\ \phi(k-s)\ \xi_f(s)$$

in place of $\xi_f(s)$ and use the trivial estimate

$$|\phi(k/2+it)|^2 \ll \left(\sum_{\nu \leq \xi} \nu^{-\frac{1}{2}}\right)^2 \ll \xi \quad ,$$

which follows from (2.3). Thus,

$$\begin{aligned} E(t;u,v) &= A(t+u,\varphi)\ \overline{A(t+v,\varphi)} + B(t+u,\varphi)\ \overline{A(t+v,\varphi)} \\ &+ A(t+u,\varphi)\ \overline{B(t+v,\varphi)} + B(t+u,\varphi)\ \overline{B(t+v,\varphi)} \\ &= E_1 + E_2 + E_3 + E_4 \quad , \end{aligned}$$

say. Assume $3 \leq \xi \leq T$ and $T/3 \leq t \leq 3T$. Then by (3.3)

$$E_2 \ll |A(t+v,\varphi)|\, T^{-\frac{1}{2}}\, \xi \log^2 T \ ,$$

$$E_3 \ll |A(t+u,\varphi)|\, T^{-\frac{1}{2}}\, \xi \log^2 T \ ,$$

$$E_4 \ll T^{-1}\, \xi^2 \log^4 T \ .$$

Thus, by Cauchy's inequality,

$$\begin{aligned} I(T;u,v) &= I_\varphi(T;u,v) + O(\xi \log^2 T\{I_\varphi^{\frac{1}{2}}(T;u,u) + I_\varphi^{\frac{1}{2}}(T,v,v)\}) \\ &+ O(\xi^2 \log^4 T) \ , \end{aligned} \tag{3.4}$$

where

$$I_\varphi(T;u,v) = \int_{-\infty}^{\infty} \Psi(t/T)\ A(t+u,\varphi)\ \overline{A(t+v,\varphi)}dt \quad . \tag{3.5}$$

Now let

$$\varphi = \varphi_1 + \varphi_2 \ , \qquad \varphi^* = \rho_1 + \rho_2 \tag{3.6}$$

with

$$\varphi_1(\tau) = \rho_1(\tau) = 0 \quad \text{if} \quad |\tau| \geqslant 2/3 \,,$$

(3.7)

$$\varphi_2(\tau) = \rho_2(\tau) = 0 \quad \text{if} \quad |\tau| \leqslant 1/2 \,.$$

Note that $\varphi + \varphi_2$ satisfies the conditions for the approximate functional equation. From (3.2)

$$(\varphi + \varphi_2)^*(\tau) = \varphi^*(\tau) - \varphi_2(1/\tau) \,,$$

and since

$$G(t) = A(t,\varphi) + B(t,\varphi)$$

$$= A(t,\varphi + \varphi_2) + B(t,\varphi + \varphi_2) \,,$$

we find that

$$\text{(3.8)} \qquad \gamma(t) \sum_{\ell,\mu,\nu} \frac{f(\ell)\beta_\mu \beta_\nu}{(\ell\mu\nu)^{k/2}} \left(\frac{\ell\mu}{\nu}\right)^{-it} \varphi_2\left(\frac{2\pi\ell\mu}{\nu t}\right)$$

$$= \gamma(-t) \sum_{\ell,\mu,\nu} \frac{f(\ell)\beta_\mu \beta_\nu}{(\ell\mu\nu)^{k/2}} \left(\frac{\ell\mu}{\nu}\right)^{it} \varphi_2\left(\frac{\nu t}{2\pi\ell\mu}\right) + B^*(t)$$

where by (3.3) and our limits on ξ and t,

$$B^*(t) = B(t,\varphi) - B(t,\varphi + \varphi_2)$$

$$= O(T^{-1} \xi \log^2 T) \,.$$

Equation (3.8) also holds with φ_2 replaced by ρ_2.

Now, (3.6) gives

$$\text{(3.9)} \qquad A(t+u,\varphi)\, \overline{A(t+v,\varphi)} = \sum_{i=1}^{4} f_i(u) \sum_{j=1}^{4} \overline{f_i(v)}$$

where

$$f_i(u) = \begin{cases} \gamma(t+u) \sum\limits_{\ell,\mu,\nu} \dfrac{f(\ell)\beta_\mu\beta_\nu}{(\ell\mu\nu)^{k/2}} \left(\dfrac{\ell\mu}{\nu}\right)^{-i(t+u)} \varphi_i\left(\dfrac{2\pi\ell\mu}{\nu(t+u)}\right), & i = 1,2 \\ \gamma(-t-u) \sum\limits_{\ell,\mu,\nu} \dfrac{f(\ell)\beta_\mu\beta_\nu}{(\ell\mu\nu)^{k/2}} \left(\dfrac{\ell\mu}{\nu}\right)^{i(t+u)} \rho_{i-2}\left(\dfrac{2\pi\ell\mu}{\nu(t+u)}\right), & i = 3,4. \end{cases}$$

In (3.9) we collect terms in the following way:

(3.10) $$A(t+u,\varphi)\,\overline{A(t+v,\varphi)}$$

$$= \{(f_1+f_2)(u)\,\overline{(f_1+f_2)(v)} + (f_1+f_2)(u)\,\overline{f_4(v)} + f_4(u)\,\overline{f_1(v)}\}$$

$$= \{(f_3+f_4)(u)\,\overline{(f_3+f_4)(v)} + (f_3+f_4)(u)\,\overline{f_2(v)} + f_2(u)\,\overline{f_3(v)}\}$$

$$+ \{f_1(u)\,\overline{f_3(v)} + f_3(u)\,\overline{f_1(v)}\}$$

$$= S_1 + S_3 + S_3 ,$$

say. In the terms involving f_2 and f_4 individually we will use (3.8). But first (3.7), (2.3) and (1.1c) imply

(3.11) $$f_i(u) \ll t^{\frac{1}{2}} \xi \log \xi , \qquad i = 1, 2, 3, 4 .$$

Furthermore, Stirling's formula yields

(3.12) $$\gamma(t+u) = a e^{-it(1-\log t)} (2\pi)^{-it} \left(\frac{t}{2\pi}\right)^{iu} \left\{1 + \frac{i(u^2+c)}{2t}\right\} + O(t^{-2})$$

where a is a complex number of modulus 1, c is a real absolute constant and the error is uniform in $0 \leqslant u \leqslant 1$. Thus by (3.8) and (3.10) - (3.10), S_1 takes the form

(3.13)

$$S_1 = \left\{1+\frac{i(u^2-v^2)}{2t}\right\} \left(\frac{t}{2\pi}\right)^{i(u-v)} \sum_{\substack{\ell,m \\ \kappa,\lambda,\mu,\nu}} \frac{f(\ell)f(m)\beta_\kappa\beta_\lambda\beta_\mu\beta_\nu}{(\ell m\kappa\lambda\mu\nu)^{k/2}}$$

$$\times \left(\frac{\ell\mu}{\nu}\right)^{-iu} \left(\frac{m\kappa}{\lambda}\right)^{iv} \left(\frac{m\kappa\nu}{\ell\mu\lambda}\right)^{it} \Phi\left(\frac{2\pi\ \ell\mu}{\nu(t+u)}, \frac{2\pi\ m\kappa}{\lambda(t+v)}\right)$$

$$+ \quad (f_1+f_2)(u)\ \overline{B^*(t+v)} \quad + \quad O(T^{-3/2}\xi^2\log^2 T)$$

where

$$\Phi(x,y) = \varphi(x)\ \varphi(y) + \varphi(x)\rho_2(1/y) + \rho_2(1/x)\ \varphi_1(y) \ .$$

Similarly, with

$$\Phi^*(x,y) = \varphi^*(x)\varphi^*(y) + \varphi^*(x)\varphi_2(1/y) + \varphi_2(1/x)\rho_1(y) \ ,$$

we have

(3.14)

$$S_2 = \left\{1+\frac{i(u^2-v^2)}{2t}\right\} \left(\frac{t}{2\pi}\right)^{i(u-v)} \sum_{\substack{\ell,m \\ \kappa,\lambda,\mu,\nu}} \frac{f(\ell)f(m)\beta_\kappa\beta_\lambda\beta_\mu\beta_\nu}{(\ell m\kappa\lambda\mu\nu)^{k/2}}$$

$$\times \left(\frac{\ell\mu}{\nu}\right)^{iu} \left(\frac{m\kappa}{\lambda}\right)^{-iv} \left(\frac{m\kappa\nu}{\ell\mu\lambda}\right)^{-it} \Phi^*\left(\frac{2\pi\ \ell\mu}{\nu(t+u)}, \frac{2\pi\ m\kappa}{\lambda(t+v)}\right)$$

$$+ \quad (f_3+f_4)(u)\ \overline{B^*(t+v)} + O(T^{-3/2}\ \xi^2\log^2 T) \ .$$

We also get

(3.15)

$$S_3 = 2\,\mathrm{Re}\ a^2 \left\{1+\frac{i(u^2+v^2+2c)}{2t}\right\} \left(\frac{t}{2\pi}\right)^{i(u+v)} \sum_{\substack{\ell,m \\ \kappa,\lambda,\mu,\nu}} \frac{f(\ell)f(m)\beta_\kappa\beta_\lambda\beta_\mu\beta_\nu}{(\ell m\kappa\lambda\mu\nu)^{k/2}}$$

$$\times \left(\frac{4\pi^2 m\ell\kappa\mu e^2}{\nu\lambda t^2}\right)^{-it} \left(\frac{\ell\mu}{\nu}\right)^{-iu} \left(\frac{m\kappa}{\lambda}\right)^{-iv} \varphi_1\left(\frac{2\pi\ \ell\mu}{\nu(t+u)}\right) \rho_1\left(\frac{2\pi\ m\kappa}{\lambda(t+v)}\right)$$

$$+ \quad O(T^{-3/2}\xi^2\log^2 T) \ .$$

We insert each of the expressions for S_1, S_2 and S_3 in (3.13) - (3.15) into the expression (3.10) and then into (3.5). We consider the various terms.

First, by (2.4), the 0-terms yield a maximum of

$$0(T^{-\frac{1}{2}}\xi^2\log^2 T) \quad . \tag{3.16}$$

Secondly, by (3.3), (2.4) and Cauchy's inequality,

$$\int_{-\infty}^{\infty} \Psi(t/T)(f_1 + f_2)(u) \quad \overline{B^*(t + v)}dt \tag{3.17}$$

$$\ll \left(\int_{T/3}^{3T} |(f_1 + f_2)(u)|^2 dt \right)^{\frac{1}{2}} \xi \log^2 T$$

$$\ll \left\{ \left(\sum_{\mu,\nu} \frac{1}{\mu\nu} \right) \left(\sum_{\mu,\nu} \int_{T/3}^{3T} \left| \sum_{\ell} \frac{f(\ell)}{\ell^{k/2+it}} \varphi\left(\frac{2\pi\ell\mu}{\nu t}\right) \right|^2 dt \right) \right\}^{\frac{1}{2}} \xi \log^2 T$$

$$\ll \left\{ (\log T)^2 \, T^{1+\epsilon} \, \xi^2 \right\}^{\frac{1}{2}} \xi \log^2 T$$

$$\ll T^{\frac{1}{2}+\epsilon} \xi^2 \quad .$$

The last estimate for the integral can be found (essentially) in Good [3, p 23] or [2, p 356]. Similarly

$$\int_{-\infty}^{\infty} \Psi(t/T)(f_3 + f_4)(u) \; \overline{B^*(t+v)}dt \tag{3.18}$$

$$\ll T^{\frac{1}{2}+\epsilon} \xi^2 \quad .$$

Thirdly, for S_3, put (suppressing a number of dependencies)

$$\phi_0(t) = a^2\Psi(t/T)\left\{1+\frac{i(u^2+v^2+2c)}{2t}\right\}\left(\frac{t}{2\pi}\right)^{i(u+v)}\varphi_1\left(\frac{2\pi \ell \mu}{\nu(t+u)}\right)\rho_1\left(\frac{2\pi m\kappa}{\lambda(t+v)}\right)$$

so that

$$(3.19)\qquad \int_{-\infty}^{\infty} S_3 \quad \Psi(t/T)dt$$

$$= 2\,\mathrm{Re}\sum_{\substack{\ell,m\\ \kappa,\lambda,\mu,\nu}} \frac{f(\ell)f(m)\beta_\kappa\beta_\lambda\beta_\mu\beta_\nu}{(\ell m\kappa\lambda\mu\nu)^{k/2}}\left(\frac{\ell\mu}{\nu}\right)^{-iu}\left(\frac{m\kappa}{\lambda}\right)^{-iv}$$

$$\int_{-\infty}^{\infty}\left(\frac{4\pi^2 m\ell\kappa\mu e^2}{\lambda\nu t^2}\right)^{-it}\phi_0(t)dt .$$

But $\phi_0(t) \neq 0$ implies by (3.7) that

$$\log\left(\frac{\lambda\nu t^2}{4\pi^2\ell m\kappa\mu}\right) \geqslant 2\log 3/2 > 0 .$$

If we put for $j = 1, 2, \ldots,$

$$\phi_j(t) = \frac{\partial}{\partial t}\left\{\frac{i\,\phi_{j-1}(t)}{\log((4\pi^2\ell m\kappa\mu)/(\lambda\nu t^2))}\right\}$$

we have by construction

$$\phi_j(t) \ll t^{-j} .$$

Thus the last integral in (3.19) may, by repeated integrations by parts, be majorized by T^{1-j}. Thus (2.3) and (1.1c) together with the above yield

$$\int_{-\infty}^{\infty} S_3\ \Psi(t/T)\ dt \ll T^{1-j}\left(\sum_{\frac{\ell\mu}{\nu}\leqslant 3T}|f(\ell)\beta_\mu\beta_\nu|(\ell\mu\nu)^{-k/2}\right)^2$$

(3.20)
$$\ll T^{2-j}\xi^2\log^2\xi$$
$$\ll T^{\frac{1}{2}}\xi^2 ,$$
provided $j \geqslant 2$.

Now, the calculation in (3.17) together with (3.9) and (3.5) implies that

(3.21)
$$I_\varphi(T;u,v) \ll T^{1+\epsilon}\ \xi^2 .$$

Combining (3.13), (3.14), (3.16) - (3.18), (3.20) and (3.21) into (3.4) we find for $3 \leqslant \xi \leqslant T$,

(3.22)
$$I(T;u,v) = \int_{-\infty}^{\infty} \Psi(t/T)\ S_1(t;u,v)dt + \int_{-\infty}^{\infty} \Psi(t/T)\ S_2(t;u,v)dt + O(T^{\frac{1}{2}+\epsilon}\xi^2)$$

where $S_1(t;u,v)$ and $S_2(t;u,v)$ are the first terms in (3.13) and (3.14), respectively,

Next, we organize the sums over $\ell,m,\kappa,\lambda,\mu,\nu$ as follows. For each fixed quadruple (κ,λ,μ,ν) put

(3.23)
$$q = \text{g.c.d}(\kappa\nu,\lambda\mu),\quad a = \kappa\nu/q,\quad b = \lambda\mu/q,\ Q = \lambda\nu/q,$$

and

(3.24)
$$\phi(t;u,v) = \left\{1 + \frac{i(u^2-v^2)}{2t}\right\}\ t^{i(u-v)}\Phi\left(\frac{2\pi\ell b}{Q(t+u)}\ ,\ \frac{2\pi\ ma}{Q(t+v)}\right) .$$

Then the first term in (3.22) becomes

(3.25)

$$\sum_{\kappa,\lambda,\mu,\nu} \frac{\beta_\kappa\beta_\lambda\beta_\mu\beta_\nu}{(\kappa\lambda\mu\nu)^{k/2}} b^{-iu} a^{iv} \left(\frac{Q}{2\pi}\right)^{i(u-v)} \sum_{\ell,m} \frac{f(\ell)f(m)}{(\ell m)^{k/2}} \ell^{-iu} m^{iv}$$

$$\times \int_{-\infty}^{\infty} \Psi(t/T)\left(\frac{ma}{\ell b}\right)^{it} \phi(t;u,v)dt \quad .$$

But since $T/3 \leqslant t \leqslant 3T$,

$$\frac{\partial^j}{\partial t^j}\left\{\Psi(t/T)\ \phi(t;u,v)\right\} \ll T^{-j}, \quad j = 0, 1, \ldots$$

provided $\ell b/Q \leqslant 3T$ and $ma/Q \leqslant 3T$. Otherwise this expression is identically zero by (3.1) and (2.4). Again by repeated integrations by parts, the integral in (3.25) can be majorized by

$$T^{1-j}(\log ma/\ell b)^{-j} \quad .$$

Fix a small positive number ϵ and let $V = T^{1-\epsilon}$. Then that part of the sum in (3.25) where

$$ma > \ell b(V + 1)/V$$

can be estimated using (2.3), (1.1c) and (3.23) by

$$\sum_{\kappa,\lambda,\mu,\nu} (\kappa\lambda\mu\nu)^{-\frac{1}{2}} \sum_{\ell \leqslant 3QT/b} |f(\ell)|\ell^{-k/2} \sum_{m\leqslant 3QT/a} |f(m)|m^{-k/2}\, T^{1-j}V^j$$

$$\ll T^{1-j\epsilon} \sum_{\kappa,\lambda,\mu,\nu} (\kappa\lambda\mu\nu)^{-\frac{1}{2}} (TQ/b)^{\frac{1}{2}}(TQ/a)^{\frac{1}{2}}$$

$$\ll T^{2-j\epsilon}\xi^2\log^2\xi$$

$$\ll T^{\frac{1}{2}+\epsilon}\xi^2 \quad ,$$

providing j is sufficiently large depending on ϵ .

Note that j can be taken only as large as the number of derivatives of Ψ and φ, φ_1 etc. For fixed ϵ, these functions can be appropriately chosen. We get a similar estimate for those terms when

$$ma < \ell b(V - 1)/V .$$

Thus writing $n = \ell b - ma$,

$$(3.26)\qquad \int_{-\infty}^{\infty} \Psi(t/T)\, S_1(t;u,v)dt$$

$$= \sum_{\kappa,\lambda,\mu,\nu} \frac{\beta_\kappa\beta_\lambda\beta_\mu\beta_\nu}{q^k} \left(\frac{Q}{2\pi}\right)^{i(u-v)} \sum_{\ell} \sum_{|n|\leq \frac{\ell b}{V}} \frac{f(\ell)f(\frac{\ell b+n}{a})}{(\ell b)^{k/2+iu}(\ell b+n)^{k/2-iv}}$$

$$\times \int_{-\infty}^{\infty} \Psi(t/T)\left(\frac{\ell b+n}{\ell b}\right)^{it} \phi(t;u,v)dt$$

$$+ \quad O(T^{\frac{1}{2}+\epsilon}\xi^2) \quad .$$

Now by Taylor's Theorem and (3.24),

$$\phi(t;v,u) = t^{i(u-v)}\Phi\left(\frac{2\pi\ell b}{Qt}, \frac{2\pi(\ell b+n)}{Qt}\right) + E(t,\ell b/Q)$$

where

$$E(t,R) = \begin{cases} O(t^{-1}) & , \quad |t| \geq 3R \\ 0 & , \quad |t| \leq 3R . \end{cases}$$

This last error contributes to the expression in (3.26) at most

$$T^\epsilon \xi^4 \ll T^{\frac{1}{2}+\epsilon}\xi^2$$

if $\xi \leq T^{\frac{1}{4}}$, a condition we assume throughout the sequel.

A similar calculation with $S_2(t;u,v)$ yields in combination with the above estimates

$$(3.27)\qquad I(T;u,v) = \sum_{\kappa\ \lambda\ \mu\ \nu} \frac{\beta_\kappa\beta_\lambda\beta_\mu\beta_\nu}{q^k}\left(\frac{Q}{2\pi}\right)^{i(u-v)}$$

$$\sum_{\ell}\ \sum_{|n|\leqslant \ell b/V} \frac{f(\ell)f(\frac{\ell b+n}{a})}{(\ell b)^{k/2+iu}(\ell b+n)^{k/2-iv}}$$

$$\times \int_{-\infty}^{\infty} \Psi(t/T)t^{i(u-v)} \left\{ \left(\frac{\ell b+n}{\ell b}\right)^{it} \Phi\left(\frac{2\pi\ell b}{Qt}, \frac{2\pi(\ell b+n)}{Qt}\right)\right.$$

$$\left. + \left(\frac{\ell b+n}{\ell b}\right)^{-it} \Phi^*\left(\frac{2\pi\ell b}{Qt}, \frac{2\pi(\ell b+n)}{Qt}\right)\right\} dt$$

$$+ O(T^{\frac{1}{2}+\epsilon}\xi^2) \quad .$$

Now, as in Good [3, p 27] since $\Phi(x,x) = \varphi(x)$,

$$(\ell b)^{-k/2-iu}(\ell b+n)^{-k/2+iv} \left(\frac{\ell b+n}{\ell b}\right)^{it} \Phi\left(\frac{2\pi\ell b}{Qt}, \frac{2\pi(\ell b+n)}{Qt}\right)$$

$$= (\ell b+n/2)^{-k-i(u-v)} \exp\{int/(\ell b+n/2)\}\varphi\left(\frac{2\pi(\ell b+n/2)}{Qt}\right)\left\{1 + \right.$$

$$\left. O\left(\left|\frac{n}{\ell b+n}\right| + \left(\frac{n}{\ell b+n}\right)^2 + t\left|\frac{n}{\ell b+n}\right|^3 + \left|\frac{n}{Qt}\right|\right)\right\}$$

The O-terms contribute

$$O(V^{-1} + V^{-2} + TV^{-3}) = O(V^{-1})$$

if $\ell b + |n|/2 \leqslant 3QT$ and zero otherwise. In (3.27) this amounts to at most (since $V = T^{1-\epsilon}$)

$$T^\epsilon \sum_{\kappa,\lambda,\mu,\nu} (\kappa\lambda\mu\nu)^{(k-1)/2} q^{-k}$$

$$\sum_{|n|\leqslant T^\epsilon Q}\ \sum_{\ell\leqslant 4TQ/b} \left|f(\ell)f\left(\frac{\ell b+n}{a}\right)\right| (\ell b+n/2)^{-k}$$

$$\ll T^{\epsilon}\xi^4$$
$$\ll T^{\frac{1}{2}+\epsilon}\xi^2 .$$

Combining our estimates and similar results for the term in (3.27) involving Φ^* we find

$$I(T;u,v)$$
$$= \sum_{\kappa,\lambda,\mu,\nu} \frac{\beta_\kappa\beta_\lambda\beta_\mu\beta_\nu}{q^k}\left(\frac{Q}{2\pi}\right)^{i(u-v)} \sum_{\ell} \sum_{|n|\leqslant \ell b/V} \frac{f(\ell)f(\frac{\ell b+n}{a})}{(\ell b+n/2)^{k+i(u-v)}}$$
$$\times \int_{-\infty}^{\infty} \Psi(t/T)t^{i(u-v)}\left\{\exp\{int/(\ell b+n/2)\}\varphi\left(\frac{2\pi(\ell b+n/2)}{Qt}\right)\right.$$
$$\left. + \exp\{-int/(\ell b+n/2)\}\varphi^*\left(\frac{2\pi(\ell b+n/2)}{Qt}\right)\right\} dt$$
$$+ O(T^{\frac{1}{2}+\epsilon}\xi^2)$$

$$(3.28) \quad = T \sum_{\kappa,\lambda,\mu,\nu} \frac{\beta_\kappa\beta_\lambda\beta_\mu\beta_\nu}{q^k}\left(\frac{TQ}{2\pi}\right)^{i(u-v)}$$
$$\sum_{\ell} \sum_{|n|\leqslant \ell b/V} \frac{f(\ell)f(\frac{\ell b+n}{a})}{(\ell b+n/2)^{k+i(u-v)}} g_{n/Q}\left(\frac{\ell b+n/2}{Qt}, u-v\right)$$
$$+ O(T^{\frac{1}{2}+\epsilon}\xi^2) ,$$

where for real x, δ, a,

$$g_\delta(x,a) = \int_{-\infty}^{\infty} \Psi(\tau)\tau^{ia}\{e^{i\delta\tau/x}\varphi(2\pi x/\tau) + e^{-i\delta\tau/x}\varphi^*(2\pi x/\tau)\}d\tau.$$

Note that $g_\delta(x,a) = 0$ if $x \geqslant 3$. Also, $I(T;u,v)$ depends only on the difference $a = u - v$, (with an error uniform in u and v). We make the obvious change of notation

$$I(T;u,v) = I(T;a) , \quad a = u - v .$$

Next we use the symmetry

$$\begin{pmatrix} \kappa \\ \nu \\ u \end{pmatrix} \longleftrightarrow \begin{pmatrix} \mu \\ \lambda \\ v \end{pmatrix}$$

which has the effect of interchanging a and b , sending a to $-a$ and n to $-n$. Then (3.28) can be put in the form

$$(3.29)\quad I(T;a) = T \sum_{\kappa,\lambda,\mu,\nu} \frac{\beta_\kappa\beta_\lambda\beta_\mu\beta_\nu}{q^k} \left(\frac{TQ}{2\pi}\right)^{ia}$$

$$\times \left\{ \sum_{\ell} \frac{f(\ell)f(\ell b/a)}{(\ell b)^{k+ia}}\, g_0\left(\frac{\ell b}{QT}, a\right) \right.$$

$$\left. + 2\,\mathrm{Re} \sum_{1 \leqslant n \leqslant T^\epsilon Q} \sum_{\ell} \frac{f(\ell)f\left(\frac{\ell b+n}{a}\right)}{(\ell b+n/2)^{k+ia}}\, g_{n/Q}\left(\frac{\ell b+n/2}{QT}, a\right) \right\}$$

$$+ \quad O(T^{\frac{1}{2}+\epsilon}\xi^2)\ .$$

We must now make a slight detour in the proof to develop the necessary analysis to estimate the sum $1 \leqslant n \leqslant T^\epsilon Q$ and to produce an asymptotic expression for the $n = 0$ term. Our object is to show that the sum on n yields at most a small (< 1) power of T and a power of ξ, so that this can be absorbed into the error. It is only the $n = 0$ term which yields the estimates in our Main Lemma.

We want to remark here that our use of smoothly decaying weight functions φ in the approximate functional equation and Ψ in the Main Lemma have allowed us to essentially ignore large portions of the "non-diagonal" terms ($n \neq 0$ terms). We have left at most $T^\epsilon \xi^2$ such terms to consider. There are other advantages to the use of these functions, particularly φ. These will be remarked on later.

4. Second Expression for $I(T,\alpha)$. We shall require the following notation and lemma. For $(a,b) = 1$, $n \geqslant 0$ and $\sigma > k$, put

$$D_{a,b}(s,n) = \sum_{\ell} f(\ell) f\left(\frac{\ell b+n}{a}\right) (\ell b+n/2)^{-s} .$$

If $n = 0$, since $f(\ell b/a) = 0$ unless $a|\ell$, the multiplicativity of f implies

$$(4.1) \qquad D_{a,b}(s,0) = (ab)^{-s} \sum_{\ell} f(\ell a) f(\ell b) \ell^{-s}$$

$$= (a\,b)^{-s} \sum_{A\sim a} f(aA) f(A) A^{-s} \sum_{B\sim b} f(bB) f(B) B^{-s} \sum_{(\ell,ab)} f^2(\ell)\ell^{-s}$$

$$= (ab)^{-s} P(a,s-k) P(b,s-k)\, D(s) ,$$

where

$$D(s) = \sum_{\ell=1}^{\infty} f^2(\ell)\ell^{-s} = \prod_{p} \left(\sum_{j=0}^{\infty} f^2(p^j) p^{-js} \right),$$

and

$$P(a,s-k) = \{ \sum_{A\sim a} f(aA) f(A) A^{-s} \} \{ \sum_{A\sim a} f^2(A) A^{-s} \}^{-1}$$

$$= \prod_{p^r \| a} \left\{ \sum_{j=0}^{\infty} f(p^{r+j}) f(p^j) p^{-js} \right\} \left\{ \sum_{j=0}^{\infty} f^2(p^j) p^{-js} \right\}^{-1} .$$

Here $A \sim a$ means A runs over all those integers composed only of primes from a, i.e., $\left(\prod_{p|A} p\right) | a$.

Delign's estimate (1.1b) shows that $P(a,s-k)$ converges absolutely for $\sigma > k - 1$, so that for fixed a , $P(a,s-k)$ is analytic in this region. Also for fixed s, $P(a,s-k)$ is a multiplicative function of a. We write

(4.2)
$$P(a,s-k) = \sum_{A\sim a} f(a,A)A^{-s}$$

where $f(a,A)$ satisfies

$$f(ab,AB) = f(a,A)f(b,B), \quad (a,b) = 1,\ A \sim a,\ B \sim b\ ;$$

$$f(a,1) = f(a)\ ;$$

$$f(a,A) \ll d_*(a)d_*(A)\ a^{(k-1)/2}A^{k-1}\ .$$

Here d_* is a generic symbol denoting any fixed multifold divisor function to any fixed power. Finally we have the estimate

(4.3)
$$P(a,s-k) \ll d_*(a)\ a^{(k-1)/2}$$

uniformly in $\sigma \geq k - \frac{1}{2}$.

Lemma 4.1. We have

$$D(s) \ll |t|^{1+\epsilon}\ ,$$

uniformly in $\sigma \geq k - \frac{1}{2}$, $|s - k| \geq \frac{1}{4}$. Also, in $\sigma \geq k - \frac{1}{2}$, $D(s)$ has only a simple pole at $s = k$ with residue K, say.

Furthermore,

$$|D_{a,b}(s,n)| \ll \frac{n|s|^{1+\epsilon}}{(ab)^{k/2-2}(\sigma - k + \frac{1}{4})}$$

uniformly in $n,\ a,\ b \geq 1,\ \sigma \geq k - \frac{1}{4}$.

Proof. The assertions concerning $D(s)$ are classical. They are consequences of the Phragmén-Lindelöf principle and Rankin's Theorem 3 in [10] that says

$$(2\pi)^{-2s}\Gamma(s)\Gamma(s+k-1)\ \zeta(2s-2k+2)\ D(s)$$

has singularities, in fact only simple poles, at $s = k$ and $k - 1$, and is invariant under $s \mapsto 2k - 1 - s$.

We deduce the final assertion from Theorem 2 of [4] which as a special case states that the series

$$\tilde{D}_{a,b}(s,n) = \sum_{\ell} f(\ell) f\left(\frac{\ell b+n}{a}\right) (\ell b+n)^{-s}$$

satisfies

$$\tilde{D}_{a,b}(s,n) \ll \frac{n^{\frac{1}{2}+\epsilon}|t|^{1+\epsilon}}{(ab)^{k/2-2}(\sigma-k+\frac{1}{4})} \tag{4.4}$$

uniformly in $n, a, b \geqslant 1$, $\sigma \geqslant k - \frac{1}{4}$. We refer the reader also to the example at the end of [4]. Thus, to deduce the assertion of the lemma, we simply note that

$$D_{a,b}(s,n) - \tilde{D}_{a,b}(s,n) = s \sum_{\ell} f(\ell) f\left(\frac{\ell b+n}{a}\right) \int_{\ell b+n/2}^{\ell b+n} u^{-s-1} du$$

$$\ll |s| n \sum_{\ell} \left| f(\ell) f\left(\frac{\ell b+n}{a}\right)\right| (\ell b+n)^{-\sigma-1}$$

$$\ll |s| n \left\{ b^{-\sigma-1} \sum_{\ell} f^2(\ell) \ell^{-\sigma-1} \right\}^{\frac{1}{2}} \left\{ a^{-\sigma-1} \sum_{\ell} f^2\left(\frac{\ell b+n}{a}\right) \left(\frac{\ell b+n}{a}\right)^{-\sigma-1} \right\}^{\frac{1}{2}}$$

$$\ll \frac{|s| n}{(ab)^{k/2}(\sigma-k+1)},$$

uniformly in $a, b, n \geqslant 1$, and $\sigma \geqslant k - 1$.

The proof of (4.4) involves the spectral theory of the Laplacian acting on the space $L^2(\Gamma_0(a,b)\backslash\mathcal{H})$ where $\mathcal{H}$ is the upper half-plane and

$$\Gamma_0(a,b) = \left\{ \begin{pmatrix} \alpha & \beta \\ \gamma & \delta \end{pmatrix} \in \Gamma(1) : a|\gamma,\ b|\beta \right\}. \tag{4.5}$$

This also requires estimates from the Fourier coefficients of the associated Maass wave forms which are uniform in a and b.

Next we compute the Mellin transform of $g_\delta(x,a)$, $\delta \geqslant 0$, $|a| \leqslant 2$. We have

$$(4.6) \qquad \int_0^\infty g_\delta(x,a)\, x^{s-1} dx = (2\pi)^{-s} M_\delta(s+1)\, \widetilde{M}(s+1+ia)$$

where

$$M_\delta(s) = \int_0^\infty \{e(\delta y)\varphi(\tfrac{1}{y}) + e(-\delta y)\varphi^*(\tfrac{1}{y})\} y^{-s}\, dy ,$$

$$\widetilde{M}(s) = \int_0^\infty \Psi(t)\, t^{s-1} dt .$$

From Good [3] we have

$$(4.7) \qquad M_0(s) = -\frac{1}{s-1} \int_0^\infty \varphi'(y)\, \{y^{s-1} + y^{1-s}\}\, dy$$

$$= \frac{2}{s-1} + O(|s-1|)$$

and $M_0(s+1)$ is an odd function of s. Also, for $j = 1, 2, \ldots$,

$$(4.8) \qquad |M_0(s)| \ll (|t| + 1)^{-j}, \qquad |s| \geqslant 1,$$

$$|\widetilde{M}(s)| \ll (|t| + 1)^{-j}$$

uniformly for σ in any bounded interval. Also $M_\delta(s)$ is entire if $\delta \neq 0$ and satisfies

$$|M_\delta^{(j)}(s)| \ll |t/\delta|^{\ell+1}$$

uniformly in $j \geqslant 0$, $\sigma \geqslant -\ell + \epsilon$, $\ell \geqslant 0$.

Thus for $(a,b) = 1$, $n \geqslant 0$, $c > 1$,

$$S(n) = \sum_{\ell} f(\ell) f\left(\frac{\ell b+n}{a}\right) (\ell b+n/2)^{-k-i\alpha} g_{n/Q}\left(\frac{\ell b+n/2}{QT};\alpha\right)$$

$$= \frac{1}{2\pi i}\left(\frac{QT}{2\pi}\right)^{-1-i\alpha} \int_{c-i\infty}^{c+i\infty} D_{a,b}(s+k-1,n)\, M_{n/Q}(s-i\alpha)\widetilde{M}(s) \times \left(\frac{QT}{2\pi}\right)^{s} ds \quad .$$

For $n \geqslant 1$, we can shift the path of integration to $\sigma = 4/5$, say with absolute convergence of the integral guaranteed by Lemma 4.1, (4.8) and (4.9). Since a, b, $Q \leqslant \xi^2$, we obtain

$$S(n) \ll (\xi^{48}/T)^{1/5}(ab)^{-k/2} \quad .$$

Inserting this estimate into (3.29) we obtain

$$(4.10) \qquad I(T,\alpha) = T \sum_{\kappa,\lambda,\mu,\nu} \frac{\beta_\kappa\beta_\lambda\beta_\mu\beta_\nu}{q^k}\left(\frac{TQ}{2\pi}\right)^{i\alpha} \sum_{\ell} \frac{f(\ell)f(\ell b/a)}{(\ell b)^{k+i\alpha}} g_0\left(\frac{\ell b}{QT};\alpha\right)$$

$$+ \; O(T^{4/5+\epsilon}\xi^{68/5}) \quad .$$

We remark here that a crucial role has been played by our weight function φ from the approximate functional equation. Because of its smoothness, we actually demonstrated that S(n) decreases polynomially in T, which is essential to this proof. Had we begun with a truncated form of the approximate functional equation S(n) would have the rough shape

$$\sum_{\ell \leqslant T} f(\ell) f\left(\frac{\ell b+n}{a}\right) (\ell b+n/2)^{-k}$$

which is at best O(1). This would not suffice. Also, without the explicit estimate like (4.4) provided in [4] the proof could not proceed. Complete uniformity is required. This, as noted in the introduction is the crucial step in the proof. We now continue.

The main term in (4.10) is more complicated because $D_{a,b}(s,0)$ has a pole as does $M_0(s)$. From Lemma 4.1, for $c > 1$, we have

$$S = \sum_{\ell} f(\ell)f(\ell b/a)(\ell b)^{-k-i\alpha} g_0(\ell b/QT;\alpha)$$

$$= \frac{1}{2\pi i}\left(\frac{QT}{2\pi}\right)^{-1-i\alpha}(ab)^{1-k}\int_{c-i\infty}^{c+i\infty} P(ab,s-1)D(s+k-1)M_0(s-i\alpha)\tilde{M}(s)\left(\frac{QT}{2\pi ab}\right)^{s}ds.$$

If $\alpha = 0$ the integrand has a double pole at $s = 1$, and if $\alpha \neq 0$ there are two simple poles at $s = 1$ and $s = 1 + i\alpha$.

Case 1. $\alpha \neq 0$. We shift the line of integration to $\sigma = 3/4$ so that since $\tilde{M}(1) = 1$,

$$S = K\left(\frac{QT}{2\pi}\right)^{-i\alpha}(ab)^{-k}P(ab,0)\, M_0(1 - i\alpha)$$

$$+ \frac{2}{(ab)^{k+i\alpha}} P(ab,i\alpha)\, D(k + i\alpha)\, \tilde{M}(1 + i\alpha)$$

$$+ O(T^{-1/4}(ab)^{-k/2}) .$$

Inserting this into (4.10) we get for $\alpha \neq 0$,

(4.11) $$I(T,c) = KT\, M_0(1 - i\alpha) \sum_{\kappa,\lambda,\mu,\nu} \frac{\beta_\kappa\beta_\lambda\beta_\mu\beta_\nu}{(\kappa\lambda\mu\nu)^k} q^k P(ab,0)$$

$$+ 2T\, \tilde{M}(1+i\alpha)\, D(k+i\alpha) \sum_{\kappa,\lambda,\mu,\nu} \frac{\beta_\kappa\beta_\lambda\beta_\mu\beta_\nu}{(\kappa\lambda\mu\nu)^k}\left(\frac{Tq}{2\pi\kappa\mu}\right)^{i\alpha} P(ab,i\alpha)$$

$$+ O(T^{4/5+\epsilon}\xi^{68/5}) .$$

Case 2. $\alpha = 0$. Just as above, we obtain for some constant K',

$$I(T,0) = I(T) = 2KT \sum_{\kappa,\lambda,\mu,\nu} \frac{\beta_\kappa \beta_\lambda \beta_\mu \beta_\nu}{(\kappa\lambda\mu\nu)^k} q^k\{P(ab,0)\left(K' + \log(Tq/ab)\right) + P'(ab,0)\}$$

$$+ O(T^{4/5+\epsilon}\xi^{68/5}) \; . \tag{4.12}$$

Next we simplify notation and collect our results in a lemma. Put $P(ab) = P(ab,0)$ and $P'(ab) = P'(ab,0)$. Also let

$$S(a) = \sum_{\kappa,\lambda,\mu,\nu} \frac{\beta_\kappa \beta_\lambda \beta_\mu \beta_\nu}{(\lambda\nu)^k} \left(\frac{q}{\kappa\mu}\right)^{k+ia} P(ab,ia) \; ,$$

$$S_1 = \sum_{\kappa,\lambda,\mu,\nu} \frac{\beta_\kappa \beta_\lambda \beta_\mu \beta_\nu}{(\kappa\lambda\mu\nu)^k} q^k P(ab)\log q \; ,$$

$$S_2 = \sum_{\kappa,\lambda,\mu,\nu} \frac{\beta_\kappa \beta_\lambda \beta_\mu \beta_\nu}{(\kappa\lambda\mu\nu)^k} q^k P(ab)\log \kappa, \tag{4.13}$$

$$S_3 = \sum_{\kappa,\lambda,\mu,\nu} \frac{\beta_\kappa \beta_\lambda \beta_\mu \beta_\nu}{(\kappa\lambda\mu\nu)^k} q^k P'(ab),$$

$$R(T,a) = \mathbb{R}\mathrm{e}\{(T/2\pi)^{ia} S(a)(ia)^{-1}\} = \mathbb{I}\mathrm{m}\{(T/2\pi)^{ia} S(a)/a\}.$$

Lemma 4.2. For $3 \leqslant \xi \leqslant T^{1/70}$, $0 \leqslant h \leqslant 1$,

$$I(T) = 2KT\{(K' + \log T)S(0) + S_1 - 2S_2 + S_3\} + O(T) \; .$$

Also,

$$J(T,h) = 4TK \int_0^h R(T,a)(h-a)da$$

$$+ O(Th^2 \sup_{0 \leqslant a \leqslant h} |S(a)|) + O\left(\frac{h^{3/2}T}{\sqrt{\log \xi}}\right)$$

Proof. The first expression is just a direct application of the notation in (4.13) into (4.12), plus the symmetry between κ and μ.

To justify the expression for $J(T,h)$, we note that

$$I(T,a) = \overline{I(T,-a)} + O(T^{4/5+\epsilon}\xi^{68/5}).$$

Thus,

$$(4.14) \qquad J(T,h) = \int_0^h \int_0^h I(T,u-v)\,du\,dv$$

$$= 2\int_0^h \mathrm{Re}\ I(T,a)(h-a)\,da + O(h^2 T^{4/5+\epsilon}\xi^{68/5}).$$

The error term here is certainly majorized by

$$h^{3/2}T\ (\log \xi)^{-\frac{1}{2}},$$

by our limits on h and ξ.

Now $\mathrm{Re}\ M_0(1-ia) = 0$ since from (4.7), $M_0(1-ia)$ is an odd function of a. Secondly,

$$D(k+ia)\ \tilde{M}(1+ia) = \frac{K}{ia} + O(1)$$

so that

$$\mathrm{Re}\{(T/2\pi)^{ia}\ D(k+ia)\ \tilde{M}(1+ia)\ S(a)\}$$

$$= K\ \mathrm{Re}\{(T/2\pi)^{ia}S(a)(ia)^{-1}\} + O(|S(a)|)$$

$$= K\ R(T,a) + O(\sup_{0\leqslant a\leqslant h} |S(a)|).$$

Combining these facts with (4.11) into (4.14) we complete the proof of this lemma.

There remains to estimate the S-sums in (4.13) and to develop an asymptotic expression for $R(T,a)$. We do this in the next section and complete the proof of our Main Lemma (and so that of our Theorem).

5. Proof of the Main Lemma. In this section we will assume all the hypothesis of the Main Lemma. That is, $\xi = T^{1/70}$ and $(\log \xi)^{-1} \leqslant h \leqslant (\log \xi)^{-9/11}$. First we write

$$S(a) = \sum_{n \leqslant \xi^2} \sum_{d|n} \mu(d) \left(\frac{n}{d}\right)^{k+ia} g_a^2(d,n) \tag{5.1}$$

where for $d|n$

$$g_a(d,n) = \sum_{\lambda\mu \equiv 0(n)} \frac{\beta_\lambda \beta_\mu}{\lambda^{k+ia}\mu^k} P(\lambda\mu d/n, ia) . \tag{5.2}$$

It is easy to show from (4.2) that for $r = 1, 2, 3, 4$

$$\mathbb{R}\, P^{(r)}(\lambda, ia) \ll d_*(\lambda)\lambda^{(k-1)/2}, \tag{5.3}$$

$$\mathbb{I}m\, P^{(r)}(\lambda, ia) \ll a d_*(\lambda)\lambda^{(k-1)/2},$$

uniformly in $0 \leqslant a \leqslant 1$. Thus, expanding $P(\lambda, ia)$ in a Taylor series about zero, we find

$$g_a(d,n) = I_0(a) + iaI_1(a) - a^2 I_2(a) + O(a^3 d(n) K(d,n)) \tag{5.4}$$

where we write

$$K(d,n) = d_*(d) \frac{G_n}{n} \left(\frac{d}{n}\right)^{(k-1)/2}$$

and with $P^{(r)}(\lambda) = P^{(r)}(\lambda, 0)$

$$I_r(a) = \sum_{\lambda\mu \equiv 0(n)} \frac{\beta_\lambda \beta_\mu}{\lambda^{k+ia}\mu^k} P^{(r)}(\lambda\mu d/n) ,$$

and G_n denotes any expression which satisfies

$$G_n \ll \prod_{p|n} (1 + p^{-3/4}) ,$$

uniformly in all parameters. (This will not necessarily be the same at each occurence.) The error estimate in (5.4) is an immediate consequence of (5.2), (5.3) and (2.3).

Lemma 5.1. For a positive integer N, $3 \leqslant X \leqslant \xi$, $0 \leqslant a \leqslant (\log \xi)^{-1/2}$ and $j = 1, 2$ put

$$h_a^{(j)}(X,N) = \sum_{(\lambda,N)=1} \frac{a_\lambda f(\lambda)(\log^+ X/\lambda)^j}{\lambda^{k+ia}}$$

Then there are positive constants c_1 and c_2 such that

$$\text{(a)} \quad h_a^{(j)}(X,N) = c_j H_N \sqrt{a}(\log X)^j + O(G_N(\log \xi)^{j-1/2})$$

where

$$H_N = \prod_{p|N}\left(1 - \frac{f^2(p)}{2p^k}\right)^{-1} \ll G_N .$$

Furthermore, if $0 \leqslant a \leqslant (\log \xi)^{-1}$ and $D \leqslant \xi/X$ then

$$\text{(b)} \quad h_a^{(j)}(X,N) \ll G_N(\log \xi)^{j-1/2} ,$$

$$\text{(c)} \quad \operatorname{Im} D^{-ia} h_a^{(1)}(X,N) \ll a\, G_N(\log \xi)^{3/2} .$$

Proof. The following identity holds in $\sigma > 0$:

$$\sum_{(\lambda,N)=1} a_\lambda f(\lambda)\, \lambda^{-k-s} = H_N(s)\, H(s)\, D(s+k)^{-1/2}$$

where

$$H_N(s) = \prod_{p|N}\left(1 - \frac{f^2(p)}{2p^{s+k}}\right)^{-1},$$

$$H(s) = \prod_p \left(1 - \frac{f^2(p)}{2p^{s+k}}\right)\left(1 + \frac{f^2(p)}{p^{s+k}} + \frac{f^2(p^2)}{p^{2s+2k}} + \dots\right)^{1/2},$$

and $H(s)$ converges absolutely in $\sigma > -\frac{1}{2}$. Note that $H_N = H_N(0)$. Lemma 4.1 together with Selberg's techniques from his proof of Lemma 12, p 14 [11], easily show that

$$h_a^{(j)}(X,N) = c_j' H_N(ia) H(ia) \sqrt{a}(\log X)^j + O(G_N(\log \xi)^{j-\frac{1}{2}}) .$$

But expanding $H_N(ia) H(ia) = H_N + O(aG_N)$ by Taylor's Theorem, and noting that

$$a^{3/2}(\log X)^j \ll a^{3/2}(\log \xi)^j \ll (\log \xi)^{j-3/4} ,$$

we prove (a). Inequality (b) is then immediate. Finally, a proof of (c) can be gleaned from [11], Lemma 13, p 16.

Lemma 5.2. For $r = 0, 1, 2$ put

$$S_a^{(r)}(X,D,N) = D^{-ia} \sum_{(\lambda,N)=1} \frac{a_\lambda P^{(r)}(\lambda) \log^+ X/\lambda}{\lambda^{k+ia}} .$$

Then, under the assumption of the previous lemma,

(a) $S_a^{(r)}(X,D,N) \ll G_N(\sqrt{a} \log \xi + \sqrt{\log \xi})$,

(b) $\mathrm{Im}\, S_a^{(r)}(X,D,N) \ll a\, G_N(\log \xi)^{3/2}$.

Proof. By (4.2)

$$(5.5) \quad P^{(r)}(\lambda) = \sum_{\ell | \lambda} \mu^2(\ell)\ell^{-k} \sum_{L\sim \ell} f(\lambda,L\ell)(\log \ell L)^r L^{-k} .$$

Then since $a_\lambda = 0$ if $\mu(\lambda) = 0$, we can write $S_a^{(r)}$ in the form

$$(5.6) \quad S_a^{(r)}(X,D,N) = \sum_{(\ell,N)=1} \frac{a_\ell}{\ell^{2k}} \sum_{L\sim\ell} \frac{f(\ell,L\ell)(\log \ell L)^r}{L^k} (\ell D)^{-ia} h_a^{(1)}(X/\ell,N\ell).$$

The lemma is then an easy consequence of (4.2) and Lemma 5.1.

Lemma 5.3. For $r = 0, 1, 2$

(a) $I_r(a) \ll \left(\sqrt{\frac{a}{\log \xi}} + \frac{1}{\log \xi}\right) K(d,n)$ for $0 \leqslant a \leqslant h$,

(b) $\operatorname{Im} I_0(a) \ll a\, K(d,n)$ for $0 \leqslant a \leqslant (\log \xi)^{-1}$.

Proof. We first rewrite $I_r(a)$ to show where the sums $S_a^{(r)}$ arise. From the multiplicativity in a of $P(a,s)$ we get for $(a,b) = 1$, $(P(a) = P^{(0)}(a))$,

$$(5.7) \qquad P(ab) = P(a)P(b) , \quad P^{(1)}(ab) = P^{(1)}(a)P(b) + P(a)P^{(1)}(b) ,$$

$$P^{(2)}(ab) = P^{(2)}(a)P(b) + 2P^{(1)}(a)P^{(1)}(b) + P(a)P^{(2)}(b) .$$

Thus using the symmetry between λ and μ , we find

$$(5.8) \qquad I_r(a) = \log^{-2}\xi \sum_{j=0}^{r} \sum_{\ell,\ell_1,m,m_1} E_j(\ell,\ell_1,m,m_1)$$

$$\times\ S_a^{(j)}(\xi/\ell mm_1, \ell_1 m\, m_1, nmm_1)$$

$$\times\ S_0^{(r-j)}(\xi/\ell_1 mm_1, 1, n\, m\, m_1) ,$$

where the summations range over

$$\ell\ell_1 \sim n,\ \ell\ell_1 \equiv 0(n),\ (m,n) = 1,\quad (m_1,mn) = 1 .$$

Also, $E_j(\ell,\ell_1,m,m_1)$ is a sum of terms of the form

$$a_{\ell mm_1}\, a_{\ell_1 mm_1}\, \mu(m_1)\, P^{(a)}(\ell\ell_1 d/n)\, P^{(b)}(m)\, P^{(c)}(m_1)\, P^{(e)}(m_1)(\ell\ell_1 m^2 m_1^2)^{-k}$$

and $a + b + c + e = j$.

Thus from Lemma 5.2(a), (5.8) and (5.3)

$$I_r(a) \ll \log^{-2}\xi \left(\frac{d}{n}\right)^{(k-1)/2} G_n(\sqrt{a}(\log \xi)^{3/2} + \log \xi) \sum_{\ell,\ell_1} \frac{d_*(\ell\ell_1 d/n)}{\ell\ell_1}$$

$$\ll \left(\sqrt{\frac{a}{\log \xi}} + \frac{1}{\log \xi}\right) K(d,n) ,$$

since the last sum is bounded by

$$\frac{1}{n} \sum_{N\sim n} \frac{d_*(Nd)d(N)}{N} \leq d_*(d) \frac{1}{n} \sum_{N\sim n} \frac{d(N)}{N} = \frac{d_*(d)G_n}{n} .$$

This proves (a). Part (b) is an easy consequence of (5.8), Lemma 5.2(b) and a calculation similar to the above.

Lemma 5.4. We have

(a) $S(a) \ll h, \quad 0 \leq a \leq h$

(b) $S(0) \ll (\log \xi)^{-1}$.

Proof. By (5.4) and Lemma 5.3(a) we get

$$g_a(d,n) \ll \left(\sqrt{\frac{h}{\log \xi}} + h^3 d(n)\right) K(d,n) .$$

Inserting this estimate in (5.1) and using the facts that

$$h \leq (\log \xi)^{-9/11} ,$$

$$\sum_{n\leq \xi} d(n)G_n/n \ll \log^2 \xi ,$$

$$\sum_{n\leq \xi} d^2(n)G_n/n \ll \log^4 \xi ,$$

we complete the proof of part (a). For part (b), we simply note that Lemma 5.3(a) (with $a = 0$) implies

(5.9) $$g_0(d,n) = I_0(0) \ll (\log \xi)^{-1} K(d,n) .$$

which in (5.11) yields the result.

Lemma 5.5. With the notation of (4.13),

(a) $S_1 \ll 1$,

(b) $S_2 \ll 1$,

(c) $S_3 \ll (\log \xi)^{-1}$.

Proof. The proof of (a) follows from (5.9) and the representation

$$S1 = \sum_{n \leq \xi^2} \sum_{d|n} \mu(d)\left(\frac{n}{d}\right)^k \log n/d \; g_0^2(d,n) .$$

For (b) we proceed as follows. By Lemma 5.4(b),

$$S_2 \ll |S(0)\log \xi - S_2| + |S(0)|\log \xi$$

$$\ll |S(0)\log \xi - S_2| + 1 .$$

The last expression in absolute values is given by

$$\sum_{n \leq \xi^2} \sum_{d|n} \mu(d)\left(\frac{n}{d}\right)^k g_0(d,n)g^*(d,n)$$

where

$$g^*(d,n) = \sum_{\lambda\mu \equiv 0(n)} \frac{\beta_\lambda\beta_\mu \log^+\xi/\lambda \cdot P(\lambda\mu d/n)}{(\lambda\mu)^k} .$$

If we decompose $g^*(d,n)$ as we did for $I_0(0)$ in the proof of Lemma 5.3 and if we use Lemma 5.1 with $r = 1$ and $r = 2$, we see that

$$g^*(d,n) \ll K(d,n) \ ,$$

from which (b) follows.

Using (5.3), (5.7) and (5.9), we can write

$$S_3 = \sum_{n<\xi^2} \sum_{d|n} \mu(d)\left(\frac{n}{d}\right)^k g_0(d,n)g'(d,n) + O((\log \xi)^{-1}) \ ,$$

where $g'(d,n)$ has a form similar to $I_1(0)$. This can be estimated exactly as before, and (c) then follows.

Our next step is to obtain an asymptotic formula for $R(T,a)$. With this and the above lemmas, the proof of the Main Lemma will be straightforward.

First by Lemma 5.3 and (5.4),

$$g_a(d,n) = I_0(a) + O\left(a\sqrt{\frac{h}{\log \xi}}\, K(d,n) + a^3 d(n)K(d,n)\right)$$

from which we deduce that (suppressing the subscript in I_0)

$$S(a) = \sum_{n<\xi^2} \sum_{d|n} \mu(d)\left(\frac{n}{d}\right)^{k+ia} I^2(a) + O\left(\sqrt{\frac{a}{\log \xi}}\right) .$$

Now by (5.6) and (5.8), and Lemmas 5.1(a) and 5.2(a),

$$(5.10) \qquad I(a) = \frac{c\sqrt{a}}{\log^2 \xi} H_n \sum \{\ldots.\} D^{-ia} + O\left(\frac{K(d,n)}{\log \xi}\right)$$

where the summation ranges over

$$\ell, \ell_1, m, m_1 \text{ as in (5.8)}, \ (r,N) = 1, \ R \sim r,$$

$D = \ell_1 m\, m_1\, r$, $L = \ell m\, m_1\, r$ and $\{\ldots.\}$ denotes

$$E_0(\ell,\ell_1,m,m_1)\, a_r r^{-2k} f(r,Rr) R^{-k} H_{mm_1 r} S_0^{(0)}(\xi/\ell_1 mm_1, 1, nmm_1)(\log \xi/L).$$

Note that a appears only in the first term of (5.10) in the form $\sqrt{a}\, D^{-ia}$, and $D \leqslant \xi^2$.

We consider two cases. The first case is for a in the range $(\log \xi)^{-1} < a \leqslant h \leqslant (\log \xi)^{-9/11}$. We square $I(a)$ in (5.10) by multiplying it by the identical expression with primes on all the summation terms. Collecting this in the main expression for $S(a)$ in $R(T,a)$ we find

$$(5.11)\qquad R(T,a) = \frac{c}{\log^4 \xi} \sum_{n \leqslant \xi^2} \sum_{d|n} \mu(d) \left(\frac{n}{d}\right)^k H_n^2 \Sigma\Sigma' \{\ldots\}\{\ldots.\}' \times \sin\left(a \log \frac{Tn}{2\pi dDD'}\right) + O((a \log \xi)^{-\frac{1}{2}}) \ .$$

We claim this also holds for $0 \leqslant a \leqslant (\log \xi)^{-1}$. To show this, it suffices to show

$$(5.12)\qquad R(T,a) \ll 1 \ ,$$

in this range. But from (5.1)

$$R(T,a) \leqslant \sum_{n \leqslant \xi^2} \sum_{d|n} \left(\frac{n}{d}\right)^k \left\{ \left|\frac{\operatorname{Im}(Tn/2\pi d)^{ia}}{a}\right| \, |g_a(d,n)|^2 + 2\, g_a(d,n)| \, \left|\frac{\operatorname{Im} g_a(d,n)}{a}\right| \right\}.$$

In the first term we apply (5.4) and Lemma 5.3(a) with $a < (\log \xi)^{-1}$. The estimate is then

$$\ll \frac{\log T}{\log \xi} \ll 1 \ ,$$

by our choice of ξ. Lemma 5.3 shows that the second term in the above is also bounded, establishing (5.12) and so (5.11) holds for $0 \leqslant a \leqslant h$.

We can now easily complete the proof of the Main Lemma. First, Lemmas 5.4(b), 5.5 and 4.2 immediately give (A).

Lemma 5.4(a), the estimate

$$\int_0^h (h-\alpha)\sin \alpha X \, d\alpha \ll h/X$$

together with (5.11) and the fact that

$$\log \xi \ll \log(Tn/2\pi dDD') \ll \log \xi$$

give

$$J(T,h) \ll T\left(\frac{h}{\log \xi} + \frac{h^{3/2}}{\sqrt{\log \xi}}\right) \ll \frac{Th^{3/2}}{\sqrt{\log \xi}}$$

which is (B). This completes our proofs.

ACKNOWLEDGMENTS

The author wishes to dedicate this paper to Professor Henryk Iwanieč for his lectures which inspired the completion of this work.

NOTES

1 Research partially supported by NSF Grant MCS 77-18723A03 at the Institute for Advanced Study, Princeton, New Jersey, and by a Bateman Research Instructorship at Caltech, Pasadena, California.

REFERENCES

[1] P. Deligne, La conjecture de Weil, I, Inst. Hautes Etudes, Sci. Publ. Math. 43 (1974), 273-307.

[2] A. Good, Approximative Funktionalgleichungen und Mittelwertsätze für Dirichletreihen, die Spitzenformen assoziiert sind, Comment. Math. Helv. 50 (1975), 327-361.

[3] A. Good, Beitraege zur Theorie der Dirichletreihen, die

Spitzenformen zugeordnet sind, J. Number Theory, 13 (1981), 18-65.

[4] J. L. Hafner, Explicit estimates in the arithmetic theory of cusp forms and Poincaré series, Math. Ann. 264 (1983), 9-20.

[5] G. H. Hardy, Note on Ramanujan's function $\tau(n)$, Proc. Cambridge Phil. Soc. 23 (1927), 675-680.

[6] E. Hecke, Über die Bestimmung Dirichletscher Reihen durch ihre Funktionalgleichung, Math. Ann. 112 (1936), 664-699.

[7] E. Hecke, Über Modulfunktionen und die Dirichletschen Reihen mit Eulerscher Produktentwicklung, Math. Ann. 114 (1937), 1-28.

[8] C. G. Lekkerkerker, On the zeros of a class of Dirichlet series, Dissertation, Utrecht, 1955.

[9] L. J. Mordell, On Ramanujan's empirical expansions of modular functions, Proc. Cambridge Phil. Soc. 19 (1920), 117-124.

[10] R. A. Rankin, Contributions to the theory of Ramanujan's function $\tau(n)$ and similar arithmetical functions, I and II, Proc. Cambridge Phil. Soc. 35 (1939), 351-372.

[11] A. Selberg, On the zeros of Riemann's zeta-function, Skr. Norske. Vid. Akad. Oslo I, 10 (1942), 59 pp.

[12] E. C. Titchmarsh, "The Theory of the Riemann Zeta Function," Oxford Univ. Press (Clarendon), London/New York, 1951.

James Lee Hafner
Department of Mathematics C-012
University of California, San Diego
La Jolla, CA 92093

Lectures on the Linear Sieve

H. Halberstam

1. The purpose of these lectures is to give a simple and transparent account of the linear Rosser-Iwaniec sieve. I follow Motohashi's memoir prepared for the Tata Institute in 1981 in basing the account on a Fundamental Lemma, in much the same way that "Sieve Methods" presented the Jurkat-Richert method;[1] and I am greatly indebted to Professor Motohashi for letting me see early drafts of his important lectures. By insisting on working with the weaker condition (1.6) below, first introduced by Iwaniec, I was led to notice certain simplifications which seem to me to offer some new insights into the Rosser-Iwaniec mechanism and render this method more readily accessible.

As usual, let $\mathcal{A}$ denote a finite integer sequence, of cardinality $|\mathcal{A}|$, and put $\mathcal{A}_d = \{a \in \mathcal{A} : a \equiv 0 \bmod d\}$. Let $\mathcal{P}$ denote a set of distinct primes, and write

$$P(z) = \prod_{\substack{p < z \\ p \in \mathcal{P}}} p \quad (z \geqslant 2), \; P(z_1,x) = P(z)/P(z_1)$$

$$= \prod_{\substack{z_1 \leqslant p < z \\ p \in \mathcal{P}}} p, \; 2 \leqslant z_1 < z \; .$$

Our objective is to study the sifting function

$$(1.1) \qquad S(\mathcal{A},\mathcal{P},z) := |\{a \in \mathcal{A} : (a,P(z)) = 1\}| = \sum_{d|P(z)} \mu(d)|\mathcal{A}_d|$$

under rather weak conditions on $\mathcal{A}$ and $\mathcal{P}$. Before describing these note that, alternatively,

$$(1.2)\qquad S(\mathcal{A},\mathcal{P},z) = \sum_{d|P(z_1,z)} \mu(d)S(\mathcal{A}_d,\mathcal{P},z_1)\ ,\ 2 \leqslant z_1 < z\ ,$$

and this will be, eventually, the starting point of our method.

We assume that there exists a convenient approximation X to $|\mathcal{A}|$, and a non-negative, multiplicative arithmetic function $\omega(d)$ on the divisors d of $P(z)$, such that the 'remainders'

$$(1.3)\qquad R_d := |\mathcal{A}_d| - \frac{\omega(d)}{d} X$$

are small on average (in some sense) over all divisors d of $P(z)$ that are less than a certain number y. We may think of $\omega(p)/p$ as the proportion of elements of $\mathcal{A}$ that are divisible by p, and therefore we should not ever seek to sift $\mathcal{A}$ by a set $\mathcal{P}$ containing a prime p for which $\omega(p) = p$. Accordingly we require that

$$(1.4)\qquad 0 < \omega(p) < p\ ,\quad p \in \mathcal{P}\ ,$$

and for convenience we put

$$(1.5)\qquad \omega(p) = 0\quad ,\quad p \notin \mathcal{P}\ .$$

From intuitive probabilistic considerations we expect $S(\mathcal{A},\mathcal{P},z)$ to be comparable with

$$X \prod_{p<z} \left(1 - \frac{\omega(p)}{p}\right)\ .$$

Define

$$V(z) = \prod_{p<z} \left(1 - \frac{\omega(p)}{p}\right)\ ;$$

we shall assume that $\omega(p)$ is at most 1 on average in the sense that

$$(1.6)\qquad V(w)/V(x) = \prod_{w \leqslant p < x} \left(1 - \frac{\omega(p)}{p}\right)^{-1} \leqslant \frac{\log x}{\log w}\left(1 + \frac{A}{\log w}\right),$$
$$2 \leqslant w < x\ ,$$

where A is a positive number, usually (but not always) an absolute constant. We shall keep dependence of results on A explicit.

Note that, by (1.6),

$$(1.7)\qquad \sum_{w \leqslant p < x} \frac{\omega(p)}{p} \leqslant \log \frac{V(w)}{V(x)} \leqslant \log \left(\frac{\log x}{\log w}\right) + \frac{A}{\log w}\,, \qquad 2 \leqslant w < x\,.$$

If you allow me for a moment to suppress explicit mention of error terms (remembering only that the role of the important parameter y is to control the accumulation of the remainders R_d) I can state our aims quite simply: we shall see that if

$$\frac{\log z_1}{\log y} \to 0$$

then

$$S(\mathcal{A},\mathcal{P},z_1) \sim XV(z_1)$$

just as we might expect, and this under a condition much weaker than (1.6). Otherwise, for larger z, $S(\mathcal{A},\mathcal{P},z)$ is 'at most'

$$XV(z)F\left(\frac{\log y}{\log z}\right)$$

and 'at least'

$$XV(z)f\left(\frac{\log y}{\log z}\right)\,,$$

where the by now well known functions F and f have the following properties (for a further discussion see e.g., "Sieve Methods," Chapter 8.2):

$$(1.8)\qquad 0 \leqslant f(u) < 1 < F(u)\,, \quad u > 0\,,$$

$$(1.9)\qquad f(u) \uparrow 1\,,\ F(u) \downarrow 1 \quad \text{as} \quad u \to +\infty$$

(exponentially),

(1.10)
$$f(u) = \begin{cases} 0, & 0 < u \leqslant 2 , \\ 2e^{\gamma} \dfrac{\log(u-1)}{u} , & 2 \leqslant u \leqslant 4 , \end{cases}$$

(1.11)
$$F(u) = \frac{2e^{\gamma}}{u} , \quad 0 < u \leqslant 3 ,$$

where is Euler's constant; and, writing

(1.12)
$$\phi^{-} = f , \quad \phi^{+} = F ,$$

we have

(1.13)
$$u\phi^{\pm}(u) - v\phi^{\pm}(v) = \int_{v}^{u} \phi^{\mp}(t-1)dt , \quad 2 \leqslant v \leqslant u .$$

Note that, in the light of (1.10), the lower bound for $S(\mathcal{A},\mathcal{P},z)$ cited earlier is true but worthless of $z \geqslant y^{1/2}$. Note also that one should think of the parameters z_1, z, y chosen ultimately as functions of X, and of X as tending to $+\infty$.

Our main result is stated at the end of Section 6.

2. For $n > 1$, let $p(n)$ denote the least, and $q(n)$ the largest, prime factor of n. Define $p(1) = +\infty$ and $q(1) = 1$.

Let $\chi(\cdot)$ be a function defined on the set $\mathcal{D}_0 = \mathcal{D}_0(z)$ of all positive integer divisors of $P(z)$, and require that

(2.1)
$$\chi(1) = 1 .$$

Define

(2.2)
$$\bar{\chi}(n) = \begin{cases} 0 , & n = 1 , \\ \chi\left(\dfrac{n}{p(n)}\right) - \chi(n) , & 1 < n \mid P(z) . \end{cases}$$

LEMMA 2.1. For every divisor n of $P(z)$,

$$1 = \mathcal{X}(n) + \sum_{\substack{d|n \\ q(n/d) < p(d)}} \bar{\mathcal{X}}(d) .$$

Proof. Let $n = p_1 p_2 \cdots p_r$, $p_1 > \ldots > p_r$. Then the summation on the right is simply

$$\sum_{i=1}^{r} \mathcal{X}(p_1 \cdots p_{i-1}) - \mathcal{X}(p_1 \cdots p_i) = 1 - \mathcal{X}(n) .$$

Corollary. (The Basic Sieve Identity) Let $h(\cdot)$ be any arithmetic function. Then

$$\sum_{d|P(z)} \mu(d)h(d) = \sum_{d|P(z)} \mu(d)\mathcal{X}(d)h(d)$$

$$+ \sum_{m|P(z)} \mu(m)\bar{\mathcal{X}}(m) \sum_{d|P(p(m))} \mu(d)h(md).$$

From (1.1) and this Corollary we obtain at once

$$(2.3) \quad S(\mathcal{A},\mathcal{P},z) = \sum_{d|P(z)} \mu(d)\mathcal{X}(d)|\mathcal{A}_d|$$

$$+ \sum_{m|P(z)} \mu(m)\bar{\mathcal{X}}(m)S(\mathcal{A}_m,\mathcal{P},p(m)) ;$$

and (1.2) combined with Lemma 2.1 yields similarly, for $z_1 < z$,

$$(2.4) \quad S(\mathcal{A},\mathcal{P},z) = \sum_{d|P(z_1,z)} \mu(d)\mathcal{X}(d)S(\mathcal{A}_d,\mathcal{P},z_1) \quad .$$

$$+ \sum_{m|P(z_1,z)} \mu(m)\bar{\mathcal{X}}(m)S(\mathcal{A}_m,\mathcal{P},p(m)) .$$

Suppose now that $\mathcal{X}^+(\cdot)$ and $\mathcal{X}^-(\cdot)$ are choices of $\mathcal{X}$ such that

$$(2.5) \quad \mu(m)\bar{\mathcal{X}}^+(m) \leqslant 0 \text{ and } \mu(m)\bar{\mathcal{X}}^-(m) \geqslant 0 \text{ for every } m|P(z).$$

Then, by (2.3),

$$(2.6) \qquad \sum_{d|P(z)} (d)\chi^-(d)|\mathcal{A}_d| \leqslant S(\mathcal{A},\mathcal{P},z) \leqslant \sum_{d|P(z)} \mu(d)\chi^+(d)|\mathcal{A}_d| \ ,$$

and, by (2.4), with $z_1 < z$,

$$(2.7) \qquad \sum_{d\ P(z_1,z)} \mu(d)\chi^-(d)S(\mathcal{A}_d,\mathcal{P},z_1) \leqslant S(\mathcal{A},\mathcal{P},z)$$

$$\leqslant \sum_{d\ P(z_1,z)} \mu(d)\chi^+(d)S(\mathcal{A}_d,\mathcal{P},z_1) \ .$$

Example 1. Selberg's sieve. Take

$$\chi^+(d) = \mu(d) \sum_{\text{L.C.M.}\{d_1,d_2\} = d} \lambda_{d_1}\lambda_{d_2}$$

where $\lambda_1 = 1$ and otherwise the λ_d's are arbitrary. The sum on the right of (2.6) becomes

$$\sum_{a\in\mathcal{A}} \Big(\sum_{\substack{d|a \\ d|P(z)}} \lambda_d \Big)^2 \ .$$

Example 2. Brun's "pure" sieve. Let $\nu(d)$ denote the number of prime factors of d. For any choice of positive integers r and s, let

$$\chi^+(d) = \begin{cases} 1 \ , & \nu(d) \leqslant 2r \ , \\ 0 & \text{otherwise} \ , \end{cases} \qquad \text{and } \chi^-(d) = \begin{cases} 1 \ , & \nu(d) \leqslant 2s+1, \\ 0 & \text{otherwise} \ . \end{cases}$$

Note that both χ^+ and χ^- are divisor-closed in the sense that, in either case, $\chi(d) = 1$ implies $\chi(\delta) = 1$ whenever $\delta | d$.

Now restrict attention to χ's having the following structure: for $1 < n = p_1p_2\dots p_r$ $(p_1 > \dots > p_r)$ let

$$\chi(n) = \eta(p_1)\eta(p_1p_2)\dots\eta(p_1\dots p_r) \ ,$$

so that

$$\bar{\chi}(n) = \chi\Big(\frac{n}{p(n)}\Big)(1 - \eta(n)) \ .$$

By reference to (2.5), such a function χ is a χ^+ provided that the associated function η satisfies

$$0 \leqslant \eta(m) \leqslant 1 \quad \text{and} \quad \eta(m) = 1 \quad \text{if} \quad \mu(m) = 1 \;;$$

and χ is a χ^- provided

$$0 \leqslant \eta(m) \leqslant 1 \quad \text{and} \quad \eta(m) = 1 \quad \text{if} \quad \mu(m) = -1 \;.$$

In fact, the Rosser-Iwaniec choices for χ^+ and χ^- are of this type: one takes

$$\chi^+(n) = \eta^+(p_1)\eta^+(p_1p_2p_3)\,\cdots \tag{2.8}$$

where, for $\mu(m) = -1$,

$$\eta^+(m) = \begin{cases} 1 & \text{if} \quad p(m)^\beta m < y\,, \\ 0 & \text{otherwise}\;; \end{cases} \tag{2.9}$$

and

$$\chi^-(n) = \eta^-(p_1p_2)\,\eta^-(p_1p_2p_3p_4)\cdots \tag{2.10}$$

where, for $\mu(m) = 1$,

$$\eta^-(m) = \begin{cases} 1 & \text{if} \quad p(m)^\beta m < y\,, \\ 0 & \text{otherwise.} \end{cases} \tag{2.11}$$

Here β is a certain positive number $\geqslant 1$, to be chosen.

These functions χ^+ and χ^- are divisor-closed and clearly both are characteristic functions of two sub-sets $\mathcal{D}^+$ and $\mathcal{D}^-$ of $\mathcal{D}_0$. Note that $\chi^-(p) = 1$ and that $\chi^\pm(n) = 1$ implies that $n < y$. More precisely, one may readily prove

LEMMA 2.2. Suppose $n \mid P(z)$. Whenever $\chi^\pm(n) = 1$ (and $z \leqslant y^{1/2}$ in the case of χ^-),

$$\log n \leqslant \left(1 - \frac{1}{2}\left(\frac{\beta-1}{\beta+1}\right)^{\frac{1}{2}\nu(n)}\right)\log y\;.$$

Proof. To illustrate, take the case of χ^- and $\nu(n) = 2k$, so that $n = p_1 \cdots p_{2k}$ $(p_1 > \cdots > p_{2k})$. Here $\chi^-(n) = 1$ implies

$$p_{2j}^{\beta+1}\, p_{2j-1} \cdots p_1 < y \qquad (j = 1,2,\ldots,k)\;.$$

Hence

$$p_{2j+1} < p_{2j} < \left(\frac{y}{p_1 \cdots p_{2j-1}}\right)^{\frac{1}{\beta+1}}$$

and therefore

$$\frac{y}{p_1 \cdots p_{2j+1}} > \left(\frac{y}{p_1 \cdots p_{2j-1}}\right)^{\frac{\beta-1}{\beta+1}} .$$

It follows that

$$\frac{y}{p_1 \cdots p_{2k-1}} > \left(\frac{y}{p_1}\right)^{(\frac{\beta-1}{\beta+1})^{k-1}} > y^{\frac{1}{2}(\frac{\beta-1}{\beta+1})^{k-1}} \qquad (p_1 < z < y^{\frac{1}{2}})$$

by iteration; and, finally,

$$\frac{y}{p_1 \cdots p_{2k}} > \left(\frac{y}{p_1 \cdots p_{2k-1}}\right)^{\frac{\beta}{\beta+1}} > y^{\frac{1}{2}(\frac{\beta-1}{\beta+1})^{k}} .$$

Eventually we shall chose $\beta = 2$ for the linear sieve. Using this value, we may indicate in rather a 'post-hoc' manner that the Rosser-Iwaniec chose of χ^+ and χ^- is good. Take the case of χ^-. By (2.3)

$$S(\mathcal{A},\mathcal{P},z) - \sum_{d|P(z)} \mu(d)\chi^-(d)|\mathcal{A}_d|$$

$$= \sum_{\substack{1 < m|P(z) \\ \underline{\nu(m) \text{ even}}}} \chi^-\left(\frac{m}{p(m)}\right)(1 - \eta^-(m))S(\mathcal{A}_m,\mathcal{P},p(m))$$

and we want to check that we lose little by replacing the sum of non-negative terms on the right by 0. Now we have indicated that the lower estimate for $S(\mathcal{A}_m,\mathcal{P},p(m))$ will be, essentially,

$$\frac{\omega(m)}{m} XV(p(m))f\left(\frac{\log y/m}{\log p(m)}\right)$$

(y/m rather than y, because the implicit remainder terms R_{md} have to be summed over m as well as d). But f assumes a positive value only if $\log(y/m)/\log p(m) > 2$, i.e. if

$$y > p^2(m)m ,$$

and this, together with $\nu(m)$ even, implies $\eta^-(m) = 1$ and therefore $1 - \eta^-(m) = 0$!

3. To obtain a Fundamental Lemma we do not require even (1.6); it suffices to know that

$$(3.1) \qquad V(w)/V(x) \leqslant C\left(\frac{\log x}{\log w}\right)^{\kappa} \quad \text{uniformly for } 2 \leqslant w \leqslant x ,$$

for some pair of positive absolute constants $C > 1$ and κ.

Let q be a positive integer such that

$$(3.2) \qquad (q, P(z)) = 1 .$$

Then, by (2.6) and (1.3),

$$(3.3) \qquad S(\mathcal{A}_q, \mathcal{P}, z) \leqslant \frac{\omega(q)}{q} X \sum_{d|P(z)} \mu(d)\chi^+(d)\frac{\omega(d)}{d} + \sum_{d|P(z)} \mu(d)\chi^+(d)R_{qd}$$

and

$$(3.4) \qquad S(\mathcal{A}_q, \mathcal{P}, z) \geqslant \frac{\omega(q)}{q} X \sum_{d|P(z)} \mu(d)\chi^-(d)\frac{\omega(d)}{d} + \sum_{d|P(z)} \mu(d)\chi^-(d)R_{qd} ,$$

for any pair of admissible functions χ^+, χ^-. We shall use the Rosser-Iwaniec functions given by (2.8), (2.9) and (2.10), (2.11), with the parameter β at our disposal still, but to be chosen large in the end. We concentrate on the upper bound, starting from (3.3). The lower bound argument is parallel.

By Lemma 2.1, Cor., with $h(d) = \omega(d)/d$, the first sum on the right of (3.3) is

(3.5)

$$\sum_{d|P(z)} \mu(d)\chi^{+}(d)\frac{\omega(d)}{d} = \sum_{d|P(z)} \mu(d)\frac{\omega(d)}{d}$$

$$- \sum_{m|P(z)} \mu(m)\bar{\chi}^{+}(m) \sum_{d|P(p(m))} \mu(d)\frac{\omega(md)}{md}$$

$$= V(z) - \sum_{m|P(z)} \mu(m)\bar{\chi}^{+}(m)\frac{\omega(m)}{m} V(p(m))$$

$$= V(z)\{1 + \sum_{k=0}^{\infty} \sum_{\substack{m|P(z)\\ \nu(m)=2k+1}} \bar{\chi}^{+}(m)\frac{\omega(m)}{m}\frac{V(p(m))}{V(z)}\} .$$

For each divisor m of $P(z)$ with $\nu(m) = 2k+1$, $\bar{\chi}^{+}(m) = 1$ implies $\eta^{+}(m) = 0$ and hence

(3.6) $$p(m)^{\beta}m \geqslant y ;$$

it implies also $\chi^{+}(m/p(m)) = 1$ and hence, by Lemma 2.2,

(3.7) $$\log \frac{m}{p(m)} \leqslant (1 - \frac{1}{2}(\frac{\beta-1}{\beta+1})^{k})\log y .$$

Let us write

(3.8) $$y = z^{u} , \quad u \geqslant 1 ;$$

then, by (3.6), $z^{2k+1+\beta} = z^{\nu(m)+\beta} \geqslant p(m)^{\beta}m \geqslant y = z^{u}$, so that

$$k \geqslant \frac{1}{2}(u - \beta - 1) .$$

Also, (3.6) and (3.7) together imply

(3.9) $$\log p(m) \geqslant \frac{1}{2}\frac{1}{\beta+1}(\frac{\beta-1}{\beta+1})^{k} \log y ,$$

so that, by (3.1) and (3.8),

$$\frac{V(p(m))}{V(z)} \leqslant C\left(\frac{\log z}{\log p(m)}\right)^{\kappa}$$

$$= C\left(\frac{1}{u}\frac{\log y}{\log p(m)}\right)^{\kappa} \leqslant C\left(\frac{2(\beta+1)}{u}\left(\frac{\beta+1}{\beta-1}\right)^{k}\right)^{\kappa} .$$

Putting all this information into (3.5) we have

(3.10)

$$\frac{1}{V(z)} \sum_{d|P(z)} \mu(d)\chi^{+}(d) \frac{\omega(d)}{d} \leqslant 1 + \sum_{k \geqslant \frac{1}{2}(u-\beta-1)} C\left(\frac{2(\beta+1)}{u}\left(\frac{\beta+1}{\beta-1}\right)^{k}\right)^{\kappa} \frac{1}{(2k+1)!} \left(\sum_{p} \frac{\omega(p)}{p}\right)^{2k+1}$$

where the range of summation of p in the inner sum is

$$\exp\left(\frac{1}{2}\,\frac{1}{\beta+1}\left(\frac{\beta-1}{\beta+1}\right)^{k} \log y\right) \leqslant p < y^{1/u} .$$

Hence (cf the first inequality in (1.7)), by (3.1),

$$\sum_{p} \frac{\omega(p)}{p} \leqslant \log C + \kappa \log\left(\frac{2(\beta+1)}{u}\left(\frac{\beta+1}{\beta-1}\right)^{k}\right)$$

over this range. Write $\beta_0 = (\frac{\beta+1}{\beta-1})^{\frac{1}{2}\kappa}$ and use the inequality $\frac{1}{n!} < (\frac{e}{n})^{n}$ for $n \geqslant 1$. Then the sum on the right of (3.10) is at the most

(3.11) $$C(\frac{2(\beta+1)}{u})^{\kappa} \sum_{k \geqslant \frac{1}{2}(u-\beta-1)} \xi^{2k+1} ,$$

where

$$\xi = e\beta_0\{\frac{\log C + \kappa \log \frac{2(\beta+1)}{u}}{2k} + \log \beta_0\} .$$

Suppose first we know only that[2] $u \geqslant 1$. Here we choose $\beta > 1$ so that, say,

$$\beta_0 \log \beta_0 < \frac{1}{10} ,$$

and it will follow that

$$\xi < \frac{1}{e}$$

uniformly for $k \geqslant k_0(C,\kappa)$, whence the infinite series (3.11) certainly converges.

Suppose on the other hand that u is large. Here choose β so that

$$\beta + 1 = \frac{1}{3} u .$$

Then, if $u \geqslant 15$,

$$\xi < \frac{C_1}{u}, \quad C_1 = e^{1+\frac{1}{3}\kappa}\left(\frac{3}{2}\log C + 4\kappa\right),$$

and the series in (3.11) is at most

$$C\left(\frac{2}{3}\right)^{\kappa} \sum_{k \geqslant \frac{1}{3}u} \left(\frac{C_1}{u}\right)^{2k} \leqslant C\left(\frac{2}{3}\right)^{\kappa-1}\left(\frac{C_1}{u}\right)^{\frac{2}{3}u} \quad \text{if } u \geqslant 2C_1,$$

$$\leqslant \left(\frac{2}{3}\right)^{\kappa-1} C \exp\left(-\frac{u}{2}\log u\right)$$

$$\text{if also } u \geqslant C_1^4 .$$

Thus, finally, if $u \geqslant \max(15, 2C_1, C_1^4)$ where

$$C_1 = e^{1+\frac{1}{3}\kappa}\left(\frac{3}{2}\log C + 4\kappa\right),$$

(3.12) $$\sum_{d|P(z)} \mu(d)\chi^{+}(d)\frac{\omega(d)}{d} \leqslant V(z)\{1 + O(\exp(-\tfrac{1}{2}u\log u)\}$$

where the constant implicit in the O-notation depends at most on C and κ, and could, if necessary, be made explicit.

A closely similar argument for the lower bound now leads us from (3.3), (3.12) and (3.4) to

THEOREM 3.1 (Fundamental Lemma). Suppose that[3] $z^2 \leqslant y$ and let

$$u = \frac{\log y}{\log z},$$

so that $u \geqslant 2$. Let q be a positive integer such that

$$(q, P(z)) = 1 .$$

For any integer sequence $\mathcal{A}$ satisfying (3.1) there exist sequences $\{\mathcal{A}_d^+\}$, $\{\mathcal{A}_d^-\}$ assuming the values 0, 1 and -1 only and depending at most on y, C and κ, such that

$$S(\mathcal{A}_q,\mathcal{P},z) \leqslant \frac{\omega(q)}{q} XV(z)\{1+O(\exp(-\tfrac{1}{2}u\log u))\} + \sum_{\substack{d|P(z)\\ d<y}} a_d^{+}R_{qd}$$

$$\geqslant \frac{\omega(q)}{q} XV(z)\{1+O(\exp(-\tfrac{1}{2}u\log u))\} - \sum_{\substack{d|P(z)\\ d<y}} a_d^{-}R_{qd}$$

where the O-constants depend at most on C and κ.

Taking $q = 1$ in the theorem, we see that the result is interesting only if $u \to +\infty$, that is, if z is <u>very</u> small compared with y. We, however, want a result for $S(\mathcal{A},\mathcal{P},z)$ even when z is so large that u is a positive constant. For this kind of result we turn to (2.7) and apply Theorem 3.1 to each term $S(\mathcal{A}_d,\mathcal{P},z_1)$. Take, in (2.7),

$$z_1 \leqslant y^{\frac{1}{(\log\log y)^2}}, \tag{3.13}$$

and apply Theorem 3.1 (with z_1 in place of z, $y_1 = y^{\frac{1}{\log\log y}}$ in place of y and d in place of q): then u in Theorem 3.1 is

$$\frac{\log y_1}{\log z_1} \geqslant \log\log y$$

and therefore each $S(\mathcal{A}_d,\mathcal{P},z_1)$ in (2.7) satisfies

$$\left|S(\mathcal{A}_d,\mathcal{P},z_1) - \frac{\omega(d)}{d} XV(z_1)\{1+O(\exp(-\tfrac{1}{2}(\log\log y)(\log\log\log y)))\}\right|$$

$$\leqslant \sum_{\substack{m|P(z_1)\\ m<y^{1/\log\log y}}} |R_{dm}| , \quad d|P(z_1,z) .$$

In this way we get at once, subject to (3.13), and using the Rosser-Iwaniec choice of $\mathcal{X}^+$ with $\beta = 2$,

(3.14)

$$S(\mathcal{A},\mathcal{P},z) \leqslant XV(z_1) \sum_{d|P(z_1,z)} \mu(d)\chi^+(d)\frac{\omega(d)}{d}$$

$$+ \; O(XV(z_1)(\log y)^{-\frac{1}{2}\log\log\log y} \prod_{z_1 \leqslant p < z} (1+\frac{\omega(p)}{p}))$$

$$+ \sum_{\substack{n|P(z) \\ n < y^{1+\frac{1}{\log\log y}}}} |R_n| \; .$$

Of course, Theorem 3.1 may be applied because (1.6) implies (3.1) with $\kappa = 1$ and some suitable C (it suffices in (1.6) to require $A \ll \log w$).

By (1.6) the second expression on the right is

$$O(XV(z)(\log y)^{-10}) \; ,$$

say, provided y is large enough. The third expression suffers from the slight imperfection that, instead of $n < y$, we have $n < y^{1+1/\log\log y}$. We shall remove this blemish in the next section. The advantage of the above procedure is that we focus at once on the dominant term, namely

$$S^+_{z_1}(y,z) := V(z_1) \sum_{d|P(z_1,z)} \mu(d)\chi^+(d)\frac{\omega(d)}{d} \; .$$

Indeed, by (1.8) we have

(3.15)

$$S^+_{z_1}(y,z) \leqslant V(z_1) \sum_{d|P(z_1,z)} \mu(d)\chi^+(d)\frac{\omega(d)}{d}\,\phi^{(-)^{\nu(d)}}(\frac{\log y/d}{\log z_1})$$

and we shall see in Section 5 that it is not very hard to derive from (3.15) that

$$(3.16)\quad S^{+}_{z_1}(y,z) \leqslant V(z)\{F(\frac{\log y}{\log z}) + 4A\frac{\log^2 z}{\log^3 z_1}\}\ ,\quad z_1 < z < y\ ,$$

the result we require.

A closely similar analysis, suffering from the same small blemish, applies to the lower bound.

Returning for a moment to (3.15), note that

$$S^{+}_{z_1}(y,z) = V(z_1)\sum_{d|P(z_1,z)} \mu(d)\chi^{+}(d)\frac{\omega(d)}{d}\phi^{(-)^{\nu(d)}}(\frac{\log y/d}{\log z_1})$$

$$- V(z_1)\sum_{d|P(z_1,z)} \chi^{+}(d)\frac{\omega(d)}{d}$$

$$\times\ \mu(d)\{\phi^{(-)^{\nu(d)}}(\frac{\log y/d}{\log z_1}) - 1\}\ ,$$

and $\mu(d)\{\} \geqslant 0$ by (1.8). Not only that, but since $\phi^{\pm}(u) \to 1$, as $u \to +\infty$, exponentially, the loss in precision implicit in (3.15) will be seen from Section 6 below to have the nature of an admissible error term.

4. The natural[5] choice of y_1 in Theorem 3.1 (with z_1 in place of z and y_1 in place of y) is y/d for application in the terms of (2.7). Choose χ^+ and χ^- as in Section 2 again, with $\beta = 2$, and indicate this dependence on y by writing them as χ^+_y and χ^-_y. By (2.7) we have

$$(4.1)\qquad S(\mathcal{A},\mathcal{P},z) \geqslant \sum_{d|P(z_1,z)} \mu(d)\chi_y^-(d)S(\mathcal{A}_d,\mathcal{P},z_1)$$

$$= \sum_{\substack{d|P(z_1,z)\\ \nu(d)\ \text{even}}} \chi_y^-(d)S(\mathcal{A}_d,\mathcal{P},z_1)$$

$$- \sum_{\substack{d|P(z_1 z)\\ \nu(d)\ \text{odd}}} \chi_y^-(d)S(\mathcal{A}_d,\mathcal{P},z_1) \quad .$$

In the first sum on the right conditions $\chi_y^-(d) = 1$ and $\nu(d)$ even guarantee that, if $d > 1$, $p(d)^2 d < y$ (cf. (2.11) with $\beta = 2$), so that $z_1^2 < y_1$ for the lower bound part of Theorem 3.1 is fulfilled: $y_1 = y/d > p(d)^2 \geqslant z_1^2$. In the second sum, where we require the upper bound from 3.1, $\chi_y^-(d) = 1$ and $\nu(d)$ odd imply only that $p(d)d < y$ and hence that $y_1 = y/d > p(d) \geqslant z_1$; but this is sufficient for the validity of the upper bound estimate in Theorem 3.1 and accordingly we obtain

(4.2)

$$S(\mathcal{A},\mathcal{P},z) \geqslant XV(z_1) \sum_{d|P(z_1,z)} \mu(d)\chi_y^-(d) \frac{\omega(d)}{d}$$

$$+ O(XV(z_1) \sum_{d|P(z_1,z)} \chi_y^-(d) \frac{\omega(d)}{d} \exp(-\tfrac{1}{2}u_d \log u_d))$$

$$- \sum_{d|P(z_1,z)} \mu(d)\chi_y^-(d) \sum_{\substack{m|P(z_1)\\ m<y/d}} a_m^{(-)\,\nu(d)+1} R_{dm}$$

where

$$u_d = \frac{\log(y/d)}{\log z_1}$$

Let

$$V(z_1) \sum_{d|P(z_1,z)} \mu(d)\chi_y^-(d)\frac{\omega(d)}{d} = S_{z_1}^-(y,z)$$

and let us also make do with $\exp(-u_d)$ in place of $\exp(-\frac{1}{2}u_d \log u_d)$. Then (4.2) yields

$$(4.3)\quad S(\mathcal{A},\mathcal{P},z) \geqslant XS_{z_1}^-(y,z)$$

$$+ O(XV(z_1) \sum_{d|P(z_1,z)} \chi_y^-(d)\frac{\omega(d)}{d} \exp(-\frac{\log(y/d)}{\log z_1}))$$

$$- \sum_{\substack{n|P(z)\\ n<y}} \beta^-(n)R_n ,$$

where

$$\beta^-(n) = \mu(d)\chi_y^-(d)a_m^{(-)^{\nu(d)+1}} ,$$

$$n = dm \ (d|P(z_1,z) \ , \ m|P(z_1)) \ ,$$

and therefore $\beta^-(n)$ assumes only the values $0, \pm 1$.

The same sort of argument lead to

$$(4.4)\quad S(\mathcal{A},\mathcal{P},z) \leqslant XS_{z_1}^+(y,z)$$

$$+ O(XV(z_1) \sum_{d|P(z_1,z)} \chi_y^+(d)\frac{\omega(d)}{d} \exp(-\frac{\log(y/d)}{\log z_1}))$$

$$+ \sum_{\substack{n|P(z)\\ n<y}} \beta^+(n)R_n ,$$

where $\beta^+(n)$ also takes only the values $-1, 0, 1$ and, if $n = dm$ (uniquely), $d|P(z_1,z)$ and $m|P(z_1)$, then

$$\beta^+(n) = \mu(n)\chi_y^+(d)a_m^{(-)^{\nu(d)}} .$$

It remains to estimate $S^-_{z_1}(y,z)$ from below and $S^+_{z_1}(y,z)$ from above, also to estimate

$$T^{\pm}_{z_1}(y,z) := V(z_1) \sum_{d|P(z_1,z)} \chi^{\pm}_y(d)\frac{\omega(d)}{d}\exp\left(-\frac{\log(y/d)}{\log z_1}\right) .$$

These estimates will be carried out in Sections 5 and 6.

Regarding $S^-_{z_1}(y,z)$, note that, in parallel to (3.15), we have

(4.5)
$$S^-_{z_1}(y,z) \geqslant V(z_1) \sum_{d|P(z_1,z)} \mu(d)\chi^-_y(d)\frac{\omega(d)}{d}\phi^{(-)^{\nu(d)+1}}\left(\frac{\log(y/d)}{\log z_1}\right)$$

by (1.8); and in the next section we shall prove (3.16) and

$$(4.6) \qquad S^-_{z_1}(y,z) \geqslant V(z)\{f\left(\frac{\log y}{\log z}\right) - 4A\frac{\log^2 z}{\log^3 z_1}\} , \quad z_1 < z < y .$$

Note that (4.6) is worthless if $y^{1/2} \leqslant z < y$, because f takes the value 0 in this range.

5. In this section we prove (3.16) and (4.6). Some technical preparation, of a kind familiar in sieve theory, is required. Begin with the trivial identity

$$(5.1) \qquad \sum_{z_1 \leqslant p < w} \frac{\omega(p)}{p} V(p) = V(z_1) - V(w) .$$

LEMMA 5.1. Let $B(p)$ be positive, continuous and monotonic increasing. Then

$$\sum_{z_1 \leqslant p < w} \frac{\omega(p)}{p} V(p)B(p) \leqslant V(w)\log w\{\int_{z_1}^{w} \frac{B(t)}{t\log^2 t}dt + \frac{A}{\log^2 z_1} B(w)\} .$$

Apply in turn to $\phi^+(\frac{\log x/p}{\log p}) - 1$ and $1 - \phi^-(\frac{\log x/p}{\log p})$, to obtain

LEMMA 5.2. For $x \geq w^2$ and $w > z_1$,

$$\sum_{z_1 \leq p < w} \frac{\omega(p)}{p} V(p)\, \phi^+(\frac{\log x/p}{\log p}) \leq V(z_1)\, \phi^-(\frac{\log x}{\log z_1}) - V(w)\phi^-(\frac{\log x}{\log w}) + 2e^{\gamma}A\, \frac{\log w}{\log^2 z_1} V(w) .$$

For $x \geq w > z_1$,

$$\sum_{z_1 \leq p < w} \frac{\omega(p)}{p} V(p)\, \phi^-(\frac{\log x/p}{\log p}) \geq V(z_1)\, \phi^+(\frac{\log x}{\log z_1}) - V(w)\phi^+(\frac{\log x}{\log w}) - 2e^{\gamma}A\, \frac{\log w}{\log^2 z_1} V(w) .$$

We can now begin the main argument. For $2 \leq z_1 \leq z$ and $r \geq 1$ define

$$\mathcal{E}_r^- = V(z)\phi^-(\frac{\log y}{\log z})$$
$$- V(z_1) \sum_{\substack{d|P(z_1,z) \\ \nu(d) < r}} \mu(d)\mathcal{X}^-(d)\frac{\omega(d)}{d}\, \phi^{(-)^{\nu(d)+1}}(\frac{\log y/d}{\log z_1})$$
$$- (-1)^r \sum_{\substack{d|P(z_1,z) \\ \nu(d)=r}} \mathcal{X}^-(d)\, \frac{\omega(d)}{d}\, V(p(d))\phi^{(-)^{r+1}}(\frac{\log y/d}{\log p(d)}) .$$

THEOREM 5.1. For $y \geq z^2$, $z \geq z_1$ and z_1 large enough,

$$\mathcal{E}_r^- \leqslant \frac{2e^\gamma A}{\log^2 z_1}\{V(z)\log z + \sum_{\substack{d|P(z_1,z)\\ \nu(d)<r}} \frac{\omega(d)}{d} V(p(d))\log p(d)\}$$

$$< 4A \frac{\log^2 z}{\log^3 z_1} V(z) \quad .$$

Letting $r \to \infty$, this proves, using (4.5), that (4.6) is true.

In the same way, if we define, for $r \geqslant 1$,

$$\mathcal{E}_r^+ = V(z)\phi^+\left(\frac{\log y}{\log z}\right)$$

$$- V(z_1) \sum_{\substack{d|P(z_1,z)\\ \nu(d)<r}} \mu(d)\chi_y^+(d)\frac{\omega(d)}{d}\phi^{(-)^{\nu(d)}}\left(\frac{\log y/d}{\log z_1}\right)$$

$$- (-1)^r \sum_{\substack{d|P(z_1,z)\\ \nu(d)=r}} \chi_y^+(d)\frac{\omega(d)}{d}V(p(d))\phi^{(-)^r}\left(\frac{\log y/d}{\log p(d)}\right) ,$$

we obtain

THEOREM 5.2. For $y \geqslant z \geqslant z_1$ and z_1 large enough

$$\mathcal{E}_r^+ \geqslant - 4A \frac{\log^2 z}{\log^3 z_1} V(z) \quad .$$

Letting $r \to \infty$, this leads from (3.15) to (3.16).
We shall prove only Theorem 5.1 in detail.

Proof of Theorem 5.1. We have

$$\mathcal{E}_{r+1}^- - \mathcal{E}_r^- = (-1)^r \sum_{\substack{d|P(z_1,z)\\ \nu(d)=r}} \chi_y^-(d)\frac{\omega(d)}{d}\Big\{ \sum_{z_1 \leqslant p < p(d)} \frac{\omega(p)}{p}$$

$$\times\ \eta_y^-(dp)V(p)\phi^{(-)^r}\Big(\frac{\log(y/dp)}{\log p}\Big) - V(z_1)\phi^{(-)^{r+1}}\Big(\frac{\log y/d}{\log z_1}\Big)$$

$$+\ V(p(d))\,\phi^{(-)^{r+1}}\Big(\frac{\log y/d}{\log p(d)}\Big)\Big\} .$$

Now $\eta_y^-(dp) = 1$ if r is even; and if r is odd then $\eta_y^-(dp) = 1$ unless $p^3d \geqslant y$. But when r is odd and

$$p^3d \geqslant y\ ,\ \phi^{(-)^r}\Big(\frac{\log(y/dp)}{\log p}\Big) = \phi^-\Big(\frac{\log(y/d)}{\log p} - 1\Big) = 0$$

since $\dfrac{\log(y/d)}{\log p} - 1 \leqslant 2$, so that the sum in parentheses may be written more simply as

$$\sum_{z_1 \leqslant p < p(d)} \frac{\omega(p)}{p} V(p)\phi^{(-)^r}\Big(\frac{\log(y/dp)}{\log p}\Big) .$$

But Lemma 5.2 with $x = y/d$ and $w = p(d)$ applies to this sum, the side conditions $y/d \geqslant p^2(d)$ when r is even, and $y/d \geqslant p(d)$ when r is odd, being assured by the presence of $\chi_y^-(d)$. Hence, by Lemma 5.2, $(-1)^r$ times the whole expression in parentheses is at most

$$2e^{\gamma}A\,\frac{\log p(d)}{\log^2 z_1}\,V(p(d)) .$$

Summing over r we arrive at

$$\mathcal{E}_r^- \leqslant \mathcal{E}_1^- + \frac{2e^{\gamma}A}{\log^2 z_1} \sum_{\substack{d|P(z_1,z)\\ \nu(d)<r}} \frac{\omega(d)}{d} V(p(d))\log p(d) .$$

But, again by Lemma 5.2,

$$\mathcal{E}_1^- = V(z)\phi^-(\frac{\log y}{\log z}) - V(z_1)\phi^-(\frac{\log y}{\log z_1})$$

$$+ \sum_{z_1 \leqslant p < z} \frac{\omega(p)}{p} V(p)\phi^+(\frac{\log y/p}{\log p})$$

$$\leqslant 2e^{\gamma} A \frac{\log z}{\log^2 z_1} V(z) \quad ,$$

provided $y \geqslant z^2$. This proves the first inequality in Theorem 5.1.

By virtue of this inequality, and of (1.6) at the second step,

$$\mathcal{E}_r^- \leqslant \frac{2e^{\gamma} A}{\log^2 z_1} \{V(z)\log z + \sum_{r=1}^{\infty} \sum_{\substack{d|P(z_1,z) \\ \nu(d)=r}} \frac{\omega(d)}{d} V(p(d))\log p(d)\}$$

$$\leqslant 2e^{\gamma} AV(z) \frac{\log z}{\log^2 z_1} \{1 + (1+\frac{A}{\log z_1}) \sum_{r=1}^{\infty} \frac{1}{r!} (\sum_{z_1 \leqslant p < z} \frac{\omega(p)}{p})^r\}$$

$$\leqslant 2e^{\gamma} AV(z) \frac{\log z}{\log^2 z_1} (1 + \frac{A}{\log z_1}) \exp(\log(\frac{\log z}{\log z_1}) + \frac{A}{\log z_1})$$

$$< 4AV(z) \frac{\log^2 z}{\log^3 z_1}$$

provided z_1 is large enough and provided also

$$\text{(5.2)} \qquad \frac{A}{\log z_1} = o(1) \quad ,$$

as we shall suppose from now on. (As I said at the outset, in most applications A is an absolute constant.)

We have now proved (4.6), and the same method proves also (3.16).

6. Referring back to (4.3) and (4.4), it remains to estimate

$$(6.1) \quad T^{\pm}_{z_1}(y,z) = V(z_1) \sum_{d|P(z_1,z)} \chi^{\pm}_y(d)\frac{\omega(d)}{d} \exp\left(-\frac{\log(y/d)}{\log z_1}\right)$$

where $z_1 \leqslant z \leqslant y$ in the '+' case, and $z_1 \leqslant z \leqslant y^{1/2}$ in the '−' case.

We write, with B to be chosen later,

$$(6.2) \quad V(z_1)^{-1}T^{\pm} = \sum_{\nu(d) < 2B} + \sum_{\nu(d) > 2B} = \Sigma^{\pm}_1 + \Sigma^{\pm}_2 ,$$

say. By Lemma 2.2 (with $\beta = 2$)

$$\frac{\log(y/d)}{\log z_1} > \frac{1}{2.3^B}\frac{\log y}{\log z_1} ,$$

so that

$$(6.3) \quad V(z_1)\ \Sigma^{\pm}_1 < V(z)\left(\frac{V(z_1)}{V(z)}\right)^2 \exp\left(-\frac{1}{2}\,3^{-B}\frac{\log y}{\log z_1}\right) .$$

The estimation of $\Sigma^{\pm}_2$ is more complicated. In order to set up a recurrence procedure we prefer to start from

$$V(z_1)\ \Sigma^{\pm}_2 < \left(1+\frac{A}{\log z_1}\right) \sum_{\substack{d|P(z_1,z)\\ \nu(d) \geqslant 2B}} \chi^{\pm}_y(d)\frac{\omega(d)}{d} V(p(d)) \exp\left(-\frac{\log(y/d)}{\log p(d)}\right),$$

as follows easily via (1.6). Thus, by (5.2)

$$(6.4) \quad V(z_1)\ \Sigma^{\pm}_2 < 2 \sum_{r \geqslant 2B} \Sigma^{\pm}(r) ,$$

where

$$(6.5) \quad \Sigma^{\pm}(r) = \sum_{\substack{d|P(z_1,z)\\ \nu(d)=r}} \chi^{\pm}_y(d)\frac{\omega(d)}{d} V(p(d))\exp\left(-\frac{\log(y/d)}{\log p(d)}\right) .$$

Let us concentrate on the '-' case.
When $r = 2k + 1$, $\chi^-(d)$ specifies no Rosser-Iwaniec condition involving $p(d)$. Hence

$$\Sigma^{-(2k+1)} = \sum_{\substack{t \mid P(z_1,z) \\ \nu(t)=2k}} \chi_y^-(t)\frac{\omega(t)}{t} \sum_{z_1 \leqslant p < p(t)} \frac{\omega(p)}{p} V(p) \exp\left(-\frac{\log(\frac{y/t}{p})}{\log p}\right)$$

$$\leqslant e\left(1 + A\frac{\log z}{\log^2 z_1}\right) \Sigma^{-(2k)}$$

by Lemma 5.1. If A is an absolute constant, or if we strengthen (5.2) to the requirement

(6.6) $$A\frac{\log y}{\log^2 z_1} = o(1) \quad ,$$

we have

(6.7) $$\Sigma^{-(2k+1)} < 3 \quad \Sigma^{-(2k)} \quad .$$

Now

$$\Sigma^{-(2k)} = \sum_{\substack{t \mid P(z_1,z) \\ \nu(t)=2k-2}} \chi_y^-(t)\frac{\omega(t)}{t} \sum_{\substack{z_1 \leqslant p_2 < p_1 < p(t) \\ p_2^3 p_1 < y/t}} \frac{\omega(p_1)}{p_1}\frac{\omega(p_2)}{p_2}$$

$$\times \quad V(p_2)\exp\left(-\frac{\log(\frac{y/t}{p_1p_2})}{\log p_2}\right)$$

and this relation provides us with a 'two step' recurrence on the basis of the following

LEMMA 6.1. We have, for $z_1 < w < x^{1/2}$,

$$\sum_{\substack{z_1 \leqslant p_2 < p_1 < w \\ p_2^3 p_1 < x}} \frac{\omega(p_1)}{p_1} \frac{\omega(p_2)}{p_2} V(p_2)\exp\left(-\frac{\log(\frac{x}{p_1 p_2})}{\log p_2}\right)$$

$$< (1 + A\frac{\log x}{\log^2 z_1})^3 (0.6804) V(w) \exp(-\frac{\log x}{\log w}) .$$

Corollary. By (6.6),

$$\Sigma^{-(2k)} < (0.6805)\ \Sigma^{-(2k-2)} .$$

Proof Lemma 6.1.

(i) Begin by assuming that

$$w \leqslant x^{1/4}$$

In the summation conditions $p_2 < p_1$ and $p_2 < (x/p_1)^{1/3}$. But if

$$w \leqslant x^{1/4} \quad \text{then} \quad p_1 < (x/p_1)^{1/3}$$

and therefore the double sum in the Lemma may be written

$$e \sum_{z_1 \leqslant p_1 < w} \frac{\omega(p_1)}{p_1} \sum_{z_1 \leqslant p_2 < p_1} \frac{\omega(p_2)}{p_2} V(p_2)\exp(-\frac{\log x/p_2}{\log p_2})$$

$$\leqslant e \sum_{z_1 \leqslant p_1 < w} \frac{\omega(p_1)}{p_1} V(p_1)\log p_1 \{\int_{z_1}^{p_1} \frac{\exp(-\frac{\log(x/p_1)}{\log t})}{t \log^2 t}\, dt$$

$$+ Ae \frac{\exp(-\frac{\log x}{\log p_1})}{\log^2 z_1} \}$$

$$= e \sum_{z_1 \leqslant p < w} \frac{\omega(p)}{p} V(p) \log p \{ \frac{1}{\log(x/p)} (\exp(-\frac{\log x/p}{\log p}) - \exp(-\frac{\log x/p}{\log z_1})) + \frac{Ae \exp(-\frac{\log x}{\log p})}{\log^2 z_1} \}$$

$$< e^2 \sum_{z_1 \leqslant p < w} \frac{\omega(p)}{p} V(p) \frac{\log p}{\log(x/p)} \exp(-\frac{\log x}{\log p})(1 + A \frac{\log x}{\log^2 z_1})$$

by Lemma 5.1; and a second application of the same Lemma yields the bound

$$e^2(1 + A \frac{\log x}{\log^2 z_1}) V(w) \log w \{ \int_{z_1}^{w} \frac{\exp(-\frac{\log x}{\log t})}{\frac{\log x}{\log t} - 1} \frac{dt}{t \log^2 t} + \frac{\exp(-\frac{\log x}{\log w})}{\frac{\log x}{\log w} - 1} \cdot \frac{A}{\log^2 z_1} \}$$

$$\leqslant e^2(1 + A \frac{\log x}{\log^2 z_1}) V(w) \log w \{ \frac{1}{\log x} \int_{\log x/\log w}^{\log x/\log z_1} \frac{du}{u-1} + \frac{A}{\log^2 z_1} \frac{\exp(-\frac{\log x}{\log w})}{\frac{\log x}{\log w} - 1} \}$$

$$\leqslant (1 + 2A \frac{\log x}{\log^2 z_1})^2 V(w) \frac{\log w}{\log x} \cdot e \int_{\frac{\log x}{\log w} - 1}^{\infty} \frac{e^{-t}}{t} dt$$

$$= (1 + 2A \frac{\log x}{\log^2 z_1})^2 V(w) \frac{\log w}{\log x} \cdot e E_1(\frac{\log x}{\log w} - 1)$$

where

$$E_1(u) = \int_u^\infty e^{-t} \frac{dt}{t} .$$

Hence the double sum is at most

$$(1+2A\frac{\log x}{\log^2 z_1})^2 V(w)\exp(-\frac{\log x}{\log w}) \cdot e \max_{u\geqslant 4} (\frac{e^u}{u} E_1(u-1)) .$$

By elementary calculus the maximum is attained at $u = 4$, and equals

$$\frac{1}{4} e^5 E_1(3) = 0.4841383$$

from the tables. Hence the double sum is at most

$$(6.8) \qquad (1+2A\frac{\log x}{\log^2 z_1})^2 (0.4842)V(w)\exp(-\frac{\log x}{\log w}) , \quad w \leqslant x^{1/4} .$$

(ii) Suppose now that

$$x^{1/4} < w < x^{1/2} .$$

This time the double sum takes the repeated form

$$\left(\sum_{z_1 \leqslant p_1 < x^{1/4}} \; \sum_{z_1 \leqslant p_2 < p_1} + \sum_{x^{1/4} \leqslant p_1 < w} \; \sum_{z_1 \leqslant p_2 < (x/p_1)^{1/3}}\right)$$

$$\times \; e \frac{\omega(p_1)}{p_2} \frac{\omega(p_2)}{p_2} V(p_2)\exp(-\frac{\log x/p_1}{\log p_2}) ,$$

and of these the first is clearly bounded above by (6.8) with $w = x^{1/4}$, i.e. by

$$(6.9) \qquad (1 + 2A \frac{\log x}{\log^2 z_1})^2 (0.4842)V(x^{1/4})\exp(-4)$$

$$< (1 + 2A \frac{\log x}{\log^2 z_1})^3 (0.4842)4e^{-4} V(w) \frac{\log w}{\log x}$$

by (1.6).

We estimate the second repeated sum by means of two applications of Lemma 5.1, and by use of (1.6). The first application gives the upper bound

$$e \sum_{x^{1/4} \leqslant p_1 < w} \frac{\omega(p_1)}{p_1} V((\frac{x}{p_1})^{1/3}) \frac{\log (x/p_1)^{1/3}}{\log (x/p_1)} \{e^{-3} + A \frac{\log x}{\log^2 z_1} e^{-3}\}$$

$$= \frac{1}{3} e^{-2} (1 + A \frac{\log x}{\log^2 z_1}) \sum_{x^{1/4} \leqslant p < w} \frac{\omega(p)}{p} V(p) \frac{V((x/p)^{1/3})}{V(p)}$$

$$\leqslant e^{-2} (1 + A \frac{\log x}{\log^2 z_1})^2 \sum_{x^{1/4} \leqslant p < w} \frac{\omega(p)}{p} V(p) \frac{\log p}{\log (x/p)}$$

$$\leqslant e^{-2} (1 + A \frac{\log x}{\log^2 z_1})^2 V(w) \log w \{ \int_{x^{1/4}}^{w} \frac{1}{\frac{\log x}{\log t} - 1} \frac{dt}{t \log^2 t}$$

$$+ \frac{A}{\log^2 z_1} \frac{1}{\frac{\log x}{\log w} - 1} \}$$

$$\leqslant e^{-2} (1 + A \frac{\log x}{\log^2 z_1})^2 V(w) \frac{\log w}{\log x} \{ \int_{\log x/\log w}^{4} \frac{du}{u-1} + A \frac{\log x}{\log^2 z_1} \}$$

$$= (1 + A \frac{\log x}{\log^2 z_1})^2 e^{-2} V(w) \frac{\log w}{\log x} \{\log \frac{3}{\frac{\log x}{\log w} - 1} + A \frac{\log x}{\log^2 z_1} \} .$$

We have to add this to the last expression on the right of (6.9). The dominant terms are

$$e^{-2} V(w) \frac{\log w}{\log x} \{ 4e^{-2}(0.4842) + \log \frac{3}{\frac{\log x}{\log w} - 1} \}$$

$$\leqslant V(w) \exp(- \frac{\log x}{\log w}) \max_{2 \leqslant u \leqslant 4} \frac{e^{u-2}}{u} (4e^{-2} (0.4842) + \log \frac{3}{u-1}) .$$

An elementary calculation shows the maximum to occur at u = 2, and to be equal to

$$\frac{1}{2}(1.9368e^{-2} + \log 3) < 0.6804 .$$

Hence, altogether, the double sum in the Lemma is at most

$$(1 + 3A\frac{\log x}{\log^2 z_1})^3 (0.6804) V(w) \exp(-\frac{\log x}{\log w}) , \quad x^{1/4} < w < x^{1/2} .$$

This completes the proof of Lemma 6.1.

By iterating Lemma 6.1, Cor., we obtain

$$\Sigma^{-(2k)} < (0.6805)^k V(z) \exp(-\frac{\log y}{\log z})$$

$$< (0.6805)^k e^{-2} V(z) , \qquad z \leqslant y^{1/2} .$$

Hence, by (6.7),

$$\Sigma^{-(2k)} + \Sigma^{-(2k+1)} < \frac{4}{e^2}(0.6805)^k V(z) , \quad z \leqslant y^{1/2} ,$$

so that, by (6.4),

(6.10) $$V(z_1)\ \Sigma_2^- < 4(0.6805)^B V(z) , \quad z \leqslant y^{1/2} .$$

The sums $\Sigma^{+(r)}$ raise no new difficulty. We have, by (6.5),

$$\Sigma^{+(2k+1)} = \Sigma_{z_1}^{+(2k+1)}(y,z)$$

$$= \sum_{z_1 \leqslant p < \min(y^{1/3}, z)} \frac{\omega(p)}{p} \sum_{\substack{t | P(z_1,p) \\ \nu(t)=2k}} \chi_{y/p}^-(t) V(p(t)) \exp(-\frac{\log(\frac{y/p}{t})}{\log p(t)})$$

$$= \sum_{z_1 \leqslant p < \min(y^{1/3}, z)} \frac{\omega(p)}{p} \Sigma_{z_1}^{-(2k)}(\frac{y}{p}, p)$$

after classifying according to the largest prime factor of d; hence (note that $p < (y/p)^{1/2}$ in each term of this sum), from above,

$$(6.11)\quad \Sigma^{+(2k+1)} \leqslant \sum_{z_1 \leqslant p < \min(y^{1/3}, z)} \frac{\omega(p)}{p}(0.6805)^k V(p)\exp(-\frac{\log y/p}{\log p})$$

$$< 3(0.6805)^k V(z)\exp(-\frac{\log y}{\log z}) < \frac{3}{e}(0.6805)^k V(z)$$

after yet another appeal to Lemma 5.1, also to (6.6).

Finally, we relate $\Sigma^{+(2k)}$ to $\Sigma^{+(2k-1)}$ in precisely the same way (and for the same reason) that linked $\Sigma^{-(2k)}$ to $\Sigma^{-(2k+1)}$, and obtain, by Lemma 5.1,

$$\Sigma^{+(2k)} \leqslant e(1+A\frac{\log z}{\log^2 z_1})\,\Sigma^{+(2k-1)} < 4(0.6805)^{k-1}V(3)\exp(-\frac{\log y}{\log z})$$

$$< \frac{6}{e}\,(0.6805)^k V(z)\quad .$$

Combining this with (6.11), and proceeding as before, we arrive rapidly at

$$(6.12)\qquad V(z_1)\ \Sigma_2^+ < 11(0.6805)^B V(z)\ ,\quad z \leqslant y\quad .$$

Now combine (6.2), (6.3) and (6.12) to give, using (1.6) and (5.2),

$$(6.13)\qquad T_{z_1}^{\pm}(y,z) \leqslant 11\ V(z)\{(\frac{\log z}{\log z_1})^2\exp(-\frac{1}{2}\cdot 3^{-B}\frac{\log y}{\log z_1})$$

$$+\ (0.6805)^B\}$$

under the conditions attached to (6.1). Here B is still free to be chosen.

We take

$$(6.14) \qquad z_1 = \exp(\log y)^{\theta} \quad , \quad \frac{2}{3} < \theta < 1 \, ,$$

where θ is a numerical constant; and we choose B so that

$$(6.15) \qquad 3^B = \frac{1}{2} \frac{(\log y)^{1-\theta}}{\log \log y} \quad .$$

Thus, by (6.13), noting that $0.6805 < 3^{-0.35}$,

$$(6.16) \qquad T^{\pm}_{z_1}(y,z) \leqslant 11\, V(z) \left\{ \frac{1}{(\log y)^{2\theta - 1}} + 2\left(\frac{\log \log y}{\log^{1-\theta} y}\right)^{0.35} \right\}$$

$$< 22\, V(z) \frac{\log \log y}{(\log y)^{(0.35)(1-\theta)}}$$

since $\theta > \frac{2}{3}$ implies that $2\theta - 1 > (0.35)(1-\theta)$. This requirement $\theta > \frac{2}{3}$ derives from (3.16) and (4.6) where we want the term

$$A \frac{\log^2 z}{\log^3 z_1} \leqslant \frac{A}{(\log y)^{3\theta - 2}}$$

to be small. Let us fix θ by the equation

$$3\theta - 2 = (0.35)(1 - \theta) \, ,$$

so that

$$(6.17) \qquad \theta = \frac{47}{67} \quad , \quad 3\theta - 2 = \frac{7}{67} > \frac{1}{10} \quad .$$

Hence, by (6.16),

$$(6.18) \qquad T^{\pm}_{z_1}(y,z) \leqslant \frac{1}{2} V(z) (\log y)^{-1/10} \quad ,$$

where $z_1 = \exp(\log y)^{47/67}$, $z_1 \leqslant z \leqslant y$ is the '+' case and $z_1 \leqslant z \leqslant y^{1/2}$ is the '-' case. Moreover, the terms in (3.16) and (4.6) referred to earlier satisfy

$$A \frac{\log^2 z}{\log^3 z_1} \leqslant \frac{1}{2} (\log y)^{-1/10} \tag{6.19}$$

provided we require of A that it satisfy

$$A = o((\log y)^{1/250}) , \tag{6.20}$$

a requirement that more than assures the truth of (6.6).

We are now able to put all this information ((4.3) and (4.4), (3.16) and (4.6), (6.18) and (6.19)) together in a form of the main Rosser-Iwaniec result:

THEOREM 6.1. Let $\mathcal{A}$ be an integer sequence and $\mathcal{P}$ a set of primes as described in Section 1. Suppose that (1.6) holds, where A is usually an absolute constant but may be allowed to vary subject only to (6.20). Then, provided only that $z \leqslant y$,

$$S(\mathcal{A},\mathcal{P},z) \leqslant XV(z)\{F(\frac{\log y}{\log z}) + (\log y)^{-1/10}\} + \sum_{\substack{n|P(z)\\ n<y}} \beta^+(n)R_n$$

and

$$S(\mathcal{A},\mathcal{P},z) \geqslant XV(z)\ \{f(\frac{\log y}{\log z}) - (\log y)^{-1/10}\} + \sum_{\substack{n|P(z)\\ n<y}} \beta^-(n)R_n ,$$

where the coefficients $\beta^{\pm}(n)$ assume only the values -1, 0, 1 and are independent of $\mathcal{A}$.

7. One sees from the foregoing that a sieve result of the type of Theorem 6.1 is a very general result,[4] applicable to a wide range of problems. One might say that this generality becomes a weakness when any one particular application is

under consideration, and it may well happen in the long run that each particular pair $(\mathcal{A},\mathcal{P})$ will have its own Theorem 6.1 in the form of an asymptotic equality. For example, the Prime Number Theorem is a classic instance of a sieve problem, yet Theorem 6.1 has little of value to offer in this context.

Nevertheless, there is a place in Theorem 6.1 where deep information about a particular pair $(\mathcal{A},\mathcal{P})$ may be deployed, and that is, in the treatment of the remainder sums

$$(7.1) \qquad \sum_{\substack{n \mid P(z) \\ n < y}} \beta^{\pm}(n) R_n \quad .$$

We require this sum to be $o(XV(z))$, and the larger we may choose y here without disturbing the order of magnitude of the sum, the closer we shall be able to move $f(\log y/\log z)$ and $F(\log y/\log z)$ towards 1. By way of illustration, take $\mathcal{A} = \{p-2: p \leqslant x\}$ and $\mathcal{P}$ the set of all odd primes. Then $X = \ell i\ x$, $\omega(n) = n/\phi(n)$, and $R_n = E(x,n,2)$ is the remainder in the Prime Number Theorem for the arithmetic progression $2 \bmod n$. In this case Bombieri's famous theorem shows that

$$\sum_{n < y} |R_n| = o(XV(z))$$

provided y is no larger than $x^{1/2}$ (almost), and the admissibility of larger y is conjectured. About three years ago Iwaniec had the brilliant insight to perceive that the structure of the Rosser-Iwaniec χ's allows one to express the sums (7.1) in bilinear form. One knows from the pioneering work of I. M. Vinogradov, from the consequences of Vaughan's identity, and from Chen's work on almost-primes in short intervals how important this could be; and, in fact, several of the most notable recent achievements in prime number theory rest, at least in part, on Iwaniec's perception. Perhaps the most dramatic instance is provided by the work of Heath-Brown, Iwaniec and Jutila on gaps between consecutive primes.

I shall end these talks with Motohashi's elegant method for accomplishing the transformation of (7.1) to bilinear form. I depart from his exposition only by avoiding his iteration of Buchstab identities, and I improve slightly the length of the remainder sum (cf. (3.14) above) at the cost of some extra computation. To keep this at a minimum I shall suppose from now on that $A \geqslant 1$ is an absolute constant.

Let z_1 and z_2 be as before -- remember that

$$\frac{\log z_1}{\log z} \to 0 \quad .$$

Introduce the partition of the interval $[z_1, z[$:

$$\mathcal{J}:\ z_1 < z_1 z_2 < z_1 z_2^2 < \ldots < z_1 z_2^{j-1} < z_1 z_2^j < \ldots < z_1 z_2^{J-1} < z \leqslant z_1 z_2^J$$

where $z_2 < z_1$ and J is a large integer about $\log(z/z_1)/\log z_2$ in size. Denote the generic sub-interval by I. If

$$I_1 = [z_1 z_2^{j_1 - 1}, z_1 z_2^{j_1}[\quad \text{and} \quad I_2 = [z_1 z_2^{j_2 - 1}, z_1 z_2^{j_2}[\qquad \text{write}$$

$I_1 > I_2$ if $j_1 > j_2$. In a typical direct product

$$D = I_1 \otimes \ldots \otimes I_r = I_1 \ldots I_r$$

for short, with

$$I_1 > \ldots > I_r \ ,$$

write $\nu(D) = r$, $\mu(D) = (-1)^{\nu(D)}$, $q(D) = I_1$ and $p(D) = I_r$, and say that a divisor d of $P(z_1, z)$, $d = p_1 p_2 \ldots p_r$ $(p_1 > \ldots > p_r)$ belongs to D, $d \in D$, if and only if $p_j \in I_j$. Let D_0 denote the empty product and adopt the natural convention that $1 \in D_0$.

Let $\mathcal{D}$ denote the set of all direct products D and define χ on $\mathcal{D}$ in much the same way as before; require that

$$\chi(D_0) = 1 ,$$

$$\chi(D) = \eta(I_1)\eta(I_1I_2)\dots\eta(I_1\dots I_r), D = I_1\dots I_r , I_1 > \dots > I_r ,$$

$$\bar{\chi}(D) = \chi(I_1\dots I_{r-1}) - \chi(I_1\dots I_r) , D = I_1\dots I_r , I_1 > \dots > I_r ,$$

$$\bar{\chi}(D_0) = 0 .$$

Let $D_1, D_2 \in \mathcal{D}$. We write $D_2 < D_1$ only when $q(D_2) < p(D_1)$, i.e. when all the sub-intervals I making up D_2 lie to the left of all the sub-intervals of D_1 . We have at once (cf. Lemma 2.1)

(7.2) $$1 - \chi(D) = \sum_{\substack{T|D \\ q(D/T) < p(T)}} \bar{\chi}(T)$$

where $T|D$ means that T is a direct product of a sub-set of the I's making up D , and D/T is the complementary sub-collection.

One might say that if $d \in D$, then d is well-separated by the partition $\mathcal{J}$. Any divisor d of $P(z_1,z)$ that is not well-separated has nevertheless the unique representation

$$d = d_2p'pd_1 , q(d_2) < p' < p < p(d_1) ,$$

where d_1 is well-separated by $\mathcal{J}$, say $d_1 \in D_1$, p' and p are primes in a single interval $I < D_1$, and $d_2|P(z_1,p')$. Given any arithmetic function $h(\cdot)$, we have

$$(7.3) \qquad \sum_{d|P(z_1,z)} \mu(d)h(d) = \sum_{D\in\mathcal{B}} \mu(D) \sum_{d\in D} h(d) + \sum_{I} \sum_{\substack{D\in\mathcal{B}\\ I<D}} \mu(D)$$

$$\times \sum_{\substack{p'<p\\ p',p\in I}} \sum_{d\in D} \sum_{t|P(z_1,p')} \mu(t)h(tp'pd);$$

and in particular, writing $S(\mathcal{A},z)$ for $S(\mathcal{A},\mathcal{P},z)$, (cf. (1.2))

$$(7.4) \qquad S(\mathcal{A},z) = \sum_{D\in\mathcal{B}} \mu(D) \sum_{d\in D} S(\mathcal{A}_d,z_1)$$

$$+ \sum_{I} \sum_{\substack{D\in\mathcal{B}\\ I<D}} \mu(D) \sum_{\substack{p'<p\\ p',p\in I}} \sum_{d\in D} S(\mathcal{A}_{p'pd},p')$$

We now derive (cf. (2.4))

THEOREM 7.1 (Motohashi)

$$S(\mathcal{A},z) = \sum_{D\in\mathcal{B}} \mu(D)\chi(D) \sum_{d\in D} S(\mathcal{A}_d,z_1)$$

$$+ \sum_{I} \sum_{\substack{D\in\mathcal{B}\\ I<D}} \mu(D)\chi(ID) \sum_{\substack{p'<p\\ p',p\in I}} \sum_{d\in D} S(\mathcal{A}_{p'pd},p')$$

$$+ \sum_{D\in\mathcal{B}} \mu(D)\bar{\chi}(D) \sum_{d\in D} S(\mathcal{A}_d,p(d)) .$$

Proof. Apply (7.2) in the first sum on the right of (7.4), and (7.2) with ID in place of D in the second sum. Then the first two sums on the right of the statement of the theorem appear at once, and we get also two more $\bar{\chi}$ -sums, namely

$$\sum_{D \in \mathcal{D}} \mu(D) \sum_{\substack{T|D \\ q(D/T) < p(T)}} \bar{\chi}(T) \sum_{d \in D} S(\mathcal{A}_d, z_1)$$

$$+ \sum_I \sum_{\substack{D \in \mathcal{D} \\ I < D}} \mu(D) \sum_{\substack{T|ID \\ q(ID/T) < p(T)}} \bar{\chi}(T) \sum_{\substack{p' < p \\ p', p \in I}} \sum_{d \in D} S(\mathcal{A}_{p^-pd}, p').$$

In the first of these sums write D_1 for T and D_2 for D/T so that $D_2 < D_1$; and in the second sum write D_1 for T and ID_2 in place of ID/T so that $I < D_2 < D_1$ (note that I cannot be in T because of the condition $q(ID/T) < p(T)$). Then the two sums combine into

$$\sum_{D_1 \in \mathcal{D}} \mu(D_1) \bar{\chi}(D_1) \sum_{d_1 \in D_1} \{ \sum_{\substack{D_2 \in \mathcal{D} \\ D_2 < D_1}} \mu(D_2) \sum_{d_2 \in D_2} S(\mathcal{A}_{d_1 d_2}, z_1)$$

$$+ \sum_I \sum_{\substack{D_2 \in \mathcal{D} \\ I < D_2 < D_1}} \mu(D_2) \sum_{\substack{p' < p \\ p', p \in I}} \sum_{d_2 \in D_2} S(\mathcal{A}_{d_1 p^- p d_2}, p')\}$$

and by (7.4) the expression in parentheses equals $S(\mathcal{A}_{d_1}, p(d_1))$. This proves Theorem 7.1.

We now construct functions χ^+ and χ^- on $\mathcal{D}$ so as to derive upper and lower bounds for $S(\mathcal{A}, z)$ from Theorem 7.1 in the same way as we deduced (2.7) from (2.4).

For each interval $I = [z_1 z_2^{j-1}, z_1 z_2^j[$, let $i = \min(z_1 z_2^j, z)$, so that i is the right-hand end-point of I . We imitate (2.9) and (2.11): with

$$D = I_1 \ldots I_r \quad (I_1 > \ldots > I_r) \ ,$$

define

$$(7.5) \qquad \eta_x^-(D) = \begin{cases} 1, & r \text{ odd} \\ 1, & r \text{ even and } i_r^3 \ldots i_1 < x \\ 0 & \text{otherwise}, \end{cases}$$

$$\eta_x^-(D) = \begin{cases} 1, & r \text{ even} \\ 1, & r \text{ odd and } i_r^3 \ldots i_1 < x \\ 0 & \text{otherwise}, \end{cases}$$

and then define $\chi_x^-(D)$ and $\chi_x^+(D)$ accordingly (cf. (2.8) and (2.10)). From Theorem 7.1 we derive at once

$$(7.6) \qquad S(\mathcal{A},z) \leqslant \sum_{D \in \mathcal{D}} \mu(D)\chi_x^+(D) \sum_{d \in D} S(\mathcal{A}_d, z_1)$$

$$+ \sum_{I} \sum_{\substack{D \in \mathcal{D} \\ I < D \\ \mu(D)=1}} \chi_x^+(ID) \sum_{\substack{p' < p \\ p',p \in I}} \sum_{d \in D} S(\mathcal{A}_{p'pd}, z_1),$$

and

$$(7.7) \qquad S(\mathcal{A},z) \geqslant \sum_{D \in \mathcal{D}} \mu(D)\chi_x^-(D) \sum_{d \in D} S(\mathcal{A}_d, z_1)$$

$$- \sum_{I} \sum_{\substack{D \in \mathcal{D} \\ I < D \\ \mu(D)=-1}} \chi_x^-(ID) \sum_{\substack{p' < p \\ p',p \in I}} \sum_{d \in D} S(\mathcal{A}_{p'pd}, z_1);$$

here we have (a) discarded the third sum on the right of Theorem 7.1, as usual, (b) discarded from the second sum the negative (resp. positive) terms in the case of χ^+ (resp. χ^-), and (c) changed the sieving cut-off in the second sum from p' to z_1, as we may do after (b). We could, if convenient, also abandon the condition $p' < p$ in the second sum. In fact, although the second sum in (7.6) and (7.7) looks complicated, we shall see that the summation over the primes p', p of a single interval I procures a saving that reduces both the second sums (in (7.6) and (7.7)) to acceptable error terms.

It is clear what we have to do next: we substitute in (7.6) and (7.7) from the Fundamental Lemma (Theorem 3.1), but to do this we have first to choose appropriate parameters. To avoid confusion, now think of Theorem 3.1 with z_1 in place of z and y_1 (to be chosen presently) in place of y; also, work with the simpler $\exp(-u)$ in place of $\exp(-\frac{1}{2}u \log u)$, $u = \log y_1/\log z_1$.

We now have in play the parameters z_1 z_2, z, y_1 and y, as well as x; and initially we require only that they satisfy the following conditions:

$$(7.8) \qquad 2 < z_2 < z_1 < z \leqslant y^{1/2}, \; z_1 < y_1 \leqslant y^{1/4}, \; xy_1 = y,$$

$$u = \frac{\log y_1}{\log z_1} \to \infty \text{ as } y \to \infty$$

(eventually we shall choose y_1, z_1 and z_2 in terms of z). Now substitute Theorem 3.1 in (7.7) with d ($\in D$) in place of q; since $d|P(z_1,z)$, $(d,P(z_1)) = 1$. (In what follows we focus attention on the more interesting lower bound.) Concentrate first on the remainder sums that arise:

$$(7.9) \qquad \sum_{D \in \mathcal{D}} \mu(D) \chi_x^-(D) \sum_{d \in D} \sum_{\substack{m|P(z_1) \\ m < y_1}} a_m^{(-)\nu(D)+1} R_{dm}$$

$$- \sum_{I} \sum_{\substack{D \in \mathcal{D} \\ I < D \\ \mu(D)=-1}} \chi_x^-(ID) \sum_{\substack{p' < p \\ p',p \in I}} \sum_{d \in D} \sum_{\substack{m | P(z_1) \\ m < y_1}} a^+_m R_{p^- pdm} .$$

We proceed to convert these to bilinear form by means of the crucial result below. For $D = I_1 \ldots I_r$, define

$$\delta = \delta(D) = i_1 i_2 \ldots i_r , \quad \delta(D_0) = 1 .$$

LEMMA 7.1. Let $y = MN$, with $M \geqslant N \geqslant 1$, but otherwise arbitrary. Then $\chi_x^-(D) = 1$ implies that there exists a decomposition

$$D = D_1 D_2$$

satisfying

$$(7.10) \qquad y_1 \delta(D_1) < M,\ \delta(D_2) < N \ \underline{\text{or}}\ \delta(D_1) < M,\ y_1 \delta(D_2) < N .$$

Furthermore, $\chi_x^-(ID) = 1$ with $I < D$ and $\mu(D) = -1$ implies that $D = D_1 D_2$ where at least one of the following three statements is true:

$$(7.11) \begin{cases} y_1 i\delta(D_1) < M,\ i\delta(D_2) < N \ \text{or}\ i\delta(D_1) < M,\ y_1 i\delta(D_2) < N , \\ y_1 i^2\delta(D_1) < M,\ \delta(D_2) < N \ \underline{\text{or}}\ i^2\delta(D_1) < M,\ y_1 \delta(D_2) < N , \\ y_1 \delta(D_1) < M,\ i^2\delta(D_2) < N \ \underline{\text{or}}\ \delta(D_1) < M,\ y_1 i^2\delta(D_2) < N . \end{cases}$$

Proof. As usual write

$$D = I_1 \ldots I_r \qquad (I_1 > \ldots > I_r)$$

and remember that $M \geqslant y^{1/2} \geqslant z$. We argue by induction on r. When $r = 0$ and $D = D_0$, the result is obvious since $y_1 < y^{1/2} \leqslant M$. When $r = 1$, we have either

$$N > y_1 \quad \text{and} \quad M \geqslant z > i_1 ,$$

or, if $N \leqslant y_1$, then

$$M = y/N \geqslant y/y_1 \geqslant y^{1/2} y_1 \geqslant zy_1 > i_1 y_1 ,$$

by (7.8).

Suppose now that the (first part of) Lemma 7.1 has already been proved for all D's under consideration with $\nu(D) = r$, and consider $D' = DI$, $I < D$, with $\chi^{-}(ID) = 1$. If $r + 1$ is even then $i^3 i_r \ldots i_1 < x$, and if $r + 1$ is odd then $i_r^3 \ldots i_1 < x$ and, a fortiriori, $i^2 i_r \ldots i_1 < x$. In either case,

$$i^2 i_r \ldots i_1 < x , \text{ so that } y_1 i^2 i_r \ldots i_1 < xy_1 = y .$$

By the induction hypothesis there is a decomposition $D = D_1 D_2$ such that (7.10) holds, and, from above,

$$y_1 i^2 \delta(D_1) \delta(D_2) < y . \tag{7.12}$$

Suppose $y_1 \delta(D_1) < M$, $\delta(D_2) < N$. Either $y_1 i \delta(D_1) < M$, in which case we are through; or $y_1 i \delta(D_1) \geqslant M$, in which case (7.12) tells us that $i\delta(D_2) < y/M = N$, so that $y_1 \delta(D_1) < M$ and $i\delta(D_2) < N$. The other possibility from (7.10) yields the same analysis, and this completes the inductive step, and hence the proof. As for the second part of Lemma 7.1, $\mu(D) = -1$ and $\chi^{-}(ID) = 1$ together imply $\mu(ID) = 1$ and

$$i^3 i_r \ldots i_1 < x ,$$

i.e.

$$y_1 i^3 i_r \ldots i_1 < y .$$

From here on the argument runs on before, apart from some mild complication.

We return to the first sum in (7.9) and write it

$$(7.13)\qquad \sum_{\substack{D \in \mathcal{D} \\ \mu(D)=1}} \mathcal{X}_x^-(D) \sum_{d \in D} \sum_{\substack{m|P(z_1,z) \\ m < y_1}} a_m^- R_{dm}$$

$$- \sum_{\substack{D \in \mathcal{D} \\ \mu(D)=-1}} \mathcal{X}_x^-(D) \sum_{d \in D} \sum_{\substack{m|P(z_1,z) \\ m < y_1}} a_m^+ R_{dm} .$$

Each D that appears possesses a decomposition $D = D_1D_2$ satisfying (7.10), by the first part of Lemma 7.1. While this decomposition is not necessarily unique, once it has been chosen then each $d \in D$ decomposes uniquely into $d = d_1d_2$, $d_1 \in D_1$ and $d_2 \in D_2$. If the first alternative in (7.10) occurs, we write

$$k = md_1 \ , \quad \ell = d_2$$

so that $k \leqslant y_1\delta(D_1) < M$ and $\ell \leqslant \delta(D_2) < N$; there is precisely one way only in which k can be written as md_1, $m|P(z_1)$ and $d_1|P(z_1,z)$. If the other alternative in (7.10) occurs, write $k = d_1$, $\ell = md_2$. Hence the expression (7.13) is at most

$$\left(\sum_{\substack{D_1,D_2 \\ (7.10)}} 1 \right) \sup_{\underline{\rho},\underline{\sigma}} \left| \sum_{k < M} \sum_{\ell < N} \rho_k \sigma_\ell R_{k\ell} \right| ,$$

where the supremum is taken over all vectors $\underline{\rho} = \{\rho_k\}_{k < M}$, $\underline{\sigma} = \{\sigma_\ell\}_{\ell < N}$ with ρ_k, σ_ℓ taking values -1, 0 or 1. Finally, D_1 and D_2 may be chosen in at most 2^{J+2} different ways, so that the first sum in (7.9) is at most

$$(7.14) \qquad \ll 2^J \sup_{\underline{\rho},\underline{\sigma}} \Big| \sum_{k<M} \sum_{\ell<N} \rho_k \sigma_\ell R_{k\ell} \Big| .$$

The second sum in (7.9) is treated in the same way on the basis of the second part of Lemma 7.1; and altogether the expression (7.9) is

$$(7.15) \qquad \ll J2^J \sup_{\underline{\rho},\underline{\sigma}} \Big| \sum_{k<M} \sum_{\ell<N} \rho_k \sigma_\ell R_{k\ell} \Big| .$$

Iwaniec has described, in the Durham Proceedings for example, the uses to which this bilinear form structure may be put.

8. We have dealt with the remainder sums arising in (7.7) and turn now to the other terms in that inequality that emerge after substitution of the Fundamental Lemma (Theorem 3.1) with y_1, z_1 in place of y, z respectively. We have

$$S(\mathcal{A},z) \geqslant XV(z_1) \sum_{D\in\mathcal{B}} \mu(D)\chi_x^-(D) \sum_{d\in D} \frac{\omega(d)}{d}$$

$$- XV(z_1) \sum_I \sum_{\substack{D\in\mathcal{B} \\ I<D,\mu(D)=-1}} \chi_x^-(ID) \sum_{\substack{p'<p \\ p',p\in I}} \frac{\omega(p')}{p'} \frac{\omega(p)}{p} \sum_{d\in D} \frac{\omega(d)}{d}$$

$$+ O\Big(XV(z_1)\exp\Big(-\frac{\log y_1}{\log z_1}\Big) \cdot \Big(1+\sup_I \sum_{p\in I} \frac{\omega(p)}{p}\Big) \sum_{d|P(z_1,z)} \frac{\omega(d)}{d}\Big)$$

$$+ O\Big(J2^J \sup_{\underline{\rho},\underline{\sigma}} \Big| \sum_{k<M} \sum_{\ell<N} \rho_k \sigma_\ell R_{k\ell} \Big|\Big), \quad MN = y .$$

The second expression on the right is

$$\ll XV(z_1)\Big(\sup_I \sum_{p\in I} \frac{\omega(p)}{p}\Big) \sum_{d|P(z_1,z)} \frac{\omega(d)}{d},$$

and

$$\sum_{d|P(z_1,z)} \frac{\omega(d)}{d} = \prod_{z_1 \leqslant p < z} \left(1+\frac{\omega(p)}{p}\right) \leqslant \frac{V(z_1)}{V(z)} \ll \frac{\log z}{\log z_1} .$$

Also

$$\sum_{p \in I} \frac{\omega(p)}{p} \ll \log\left(\frac{\log i}{\log(i/z_2)}\right) \ll \frac{\log z_2}{\log z_1} .$$

Hence

$$(8.1) \qquad S(\mathcal{A},z) \geqslant XV(z_1) \sum_{D\in\mathcal{B}} \mu(D)\chi_x^-(D) \sum_{d \in D} \frac{\omega(d)}{d}$$

$$+ \; O\left(XV(z)\left(\frac{\log z}{\log z_1}\right)^2 \left(\exp\left(-\frac{\log y_1}{\log z_1}\right) + \frac{\log z_2}{\log z_1}\right)\right)$$

$$+ \; O\left(\frac{\log(z/z_1)}{\log z_2}\, 2^{\frac{\log(z/z_1)}{\log z_2}} \sup_{\underline{\rho},\underline{\sigma}} \left| \sum_{k<M} \sum_{\ell<N} \rho_k \sigma_\ell R_{k\ell} \right|\right) .$$

We deal with the first sum on the right of (8.1) by imitating the arguments in Section 5 (see especially Theorem 5.1). There are some slight complications, arising in part from the introduction of x in place of y in the definition of χ. Think of x as only slightly smaller than y (cf. (8.9) below).

Begin as before by defining, for $r \geqslant 1$,

$$(8.2) \qquad \mathcal{E}_r^- = V(z)\phi^-\left(\frac{\log x}{\log z}\right)$$

$$- \; V(z_1) \sum_{\substack{D \in \mathcal{B} \\ \nu(D)<T}} \mu(D)\chi_x^-(D) \sum_{d\in D} \frac{\omega(d)}{d}\, \phi^{(-)^{\nu(D)+1}}\left(\frac{\log x/d}{\log z_1}\right)$$

$$- (-1)^r \sum_{\substack{D \in \mathcal{B} \\ \nu(D)=r}} \chi_x^-(D) \sum_{d \in D} \frac{\omega(d)}{d} V(p(d))\phi^{(-)^{r+1}}\left(\frac{\log x/d}{\log p(d)}\right) .$$

By (1.8)

$$(8.3) \quad V(z_1) \sum_{D \in \mathcal{B}} \mu(D)\chi_x^-(D) \sum_{d \in D} \frac{\omega(d)}{d}$$

$$\geqslant V(z_1) \sum_{D \in \mathcal{B}} \mu(D)\chi_x^-(D) \sum_{d \in D} \frac{\omega(d)}{d}\phi^{(-)^{\nu(D)+1}}\left(\frac{\log x/d}{\log z_1}\right)$$

and therefore we obtain a lower bound for the leading term on the right of (8.1) by estimating $\mathcal{E}_r^-$, uniformly in r , from above and letting $r \to \infty$ in (8.2). By (8.2)

$$\mathcal{E}_{r+1}^- - \mathcal{E}_r^-$$

$$= \sum_{\substack{D \in \mathcal{B} \\ \nu(D)=r}} \chi_x^-(D) \sum_{d \in D} \frac{\omega(d)}{d} (-1)^r \{ \sum_{I < D} \eta_x^-(ID) \sum_{p \in I}$$

$$\frac{\omega(p)}{p} V(p)\phi^{(-)^{r+1}}\left(\frac{\log(x/pd)}{\log p}\right) - V(z_1)\phi^{(-)^{r+1}}\left(\frac{\log x/d}{\log z_1}\right)$$

$$+ V(p(d))\phi^{(-)^{r+1}}\left(\frac{\log x/d}{\log p(d)}\right) \}$$

By Lemma 5.2

$$(8.4) \quad (-1)^r\{V(p(d)\phi^{(-)^{r+1}}\left(\frac{\log x/d}{\log p(d)}\right) - V(z_1)\phi^{(-)^{r+1}}\left(\frac{\log x/d}{\log z_1}\right)$$

$$+ \sum_{z_1 \leqslant p < p(d)} \frac{\omega(p)}{p} V(p)\phi^{(-)^r}\left(\frac{\log(x/pd)}{\log p}\right) \}$$

$$\leqslant 2e^{\gamma}A\,\frac{\log p(d)}{\log^2 z_1}\,V(p(d))$$

provided that $x/d \geqslant p(d)^2$ when r is even and that $x/d \geqslant p(d)$ when r is odd; conditions that are guaranteed by $\mathcal{X}_x^-(D) = 1$ when $r \geqslant 2$. When $r = 1$, and therefore $d = p(d)$ is a prime, q say, the result is true also if $x/q \geqslant q$, i.e. $x \geqslant q^2$, but fails when $x^{1/2} < q < y^{1/2}$. In this case the expression on the left of (8.4) is at most

$$- V(q)\phi^+\left(\frac{\log x/q}{\log q}\right) + V(z_1)\phi^+\left(\frac{\log x/q}{\log z_1}\right)$$

$$- \sum_{z_1 \leqslant p < x/q} \frac{\omega(p)}{p}\,V(p)\phi^-\left(\frac{\log(x/pq)}{\log p}\right)$$

$$\leqslant V\left(\frac{x}{q}\right)\phi^+(1) - V(q)\phi^+\left(\frac{\log x/q}{\log q}\right) + 2e^{\gamma}A\,\frac{\log x/q}{\log^2 z_1}\,V\left(\frac{x}{q}\right)$$

by Lemma 5.2; since $(\log x/q)/\log q \leqslant 1$,

$$\phi^+\left(\frac{\log x/q}{\log q}\right) = 2e^{\gamma}\,\frac{\log q}{\log x/q},$$

and therefore, applying (1.5) twice, the expression under consideration is at most

$$2e^{\gamma}V(q)\left\{\frac{\log q}{\log x/q}\left(1 - \frac{A}{\log x/q}\right) - \frac{\log q}{\log x/q}\right.$$

$$\left. + \frac{A}{\log^2 z_1}\log q\left(1 + \frac{A}{\log x/q}\right)\right\}$$

$$\leqslant 2e^{\gamma}AV(q)\log q\left\{\frac{1}{\log^2 x/q} + \frac{1}{\log^2 z_1}\left(1 + \frac{A}{\log x/q}\right)\right\}.$$

Since $x/q \geqslant xy^{-1/2} = y^{1/2}y_1^{-1} \geqslant y^{1/4} > z_1$ by (7.8), this expression is at most

$$\frac{4e^{\gamma}AV(q)\log q}{\log^2 z_1},$$

(8.4) is true also when $r = 1$ provided that the factor 2 on the right is replaced by 4. Hence

$$(8.5)\qquad \mathcal{E}_{r+1}^{-} - \mathcal{E}_{r}^{-} \leqslant \sum_{\substack{d \in \mathcal{D} \\ \nu(d)=r}} \chi_x^{-}(D) \sum_{d \in D} \frac{\omega(d)}{d} (-1)^r$$

$$\times \{ \sum_{I < D} \eta_x^{-}(ID) \sum_{p \in I} \frac{\omega(p)}{p} V(p)\phi^{(-)^r}\left(\frac{\log(x/pd)}{\log p}\right)$$

$$- \sum_{z_1 \leqslant p < p(d)} \frac{\omega(p)}{p} V(p)\phi^{(-)^r}\left(\frac{\log(x/pd)}{\log p}\right) \}$$

$$+ O\left(\frac{1}{\log^2 z_1} \sum_{\substack{d \mid P(z_1,z) \\ \nu(d)=r}} \frac{\omega(d)}{d} V(p(d))\log p(d)\right).$$

The function $\eta_x^{-}(ID) = 1$ when r is even, so that when r is even

$$(-1)^r\{\ \} = - \sum_{\substack{pd \in D \\ p < p(d)}} \frac{\omega(p)}{p} V(p)\phi^{+}\left(\frac{\log(x/pd)}{\log p}\right) \leqslant 0 .$$

When r is odd, $\eta_x^{-}(ID) = 1$ unless $i^3\delta(D) \geqslant x$, and so for odd r

$$(8.6)\qquad (-1)^r\{\ \} = \sum_{\substack{pd \in D \\ p < p(d)}} \frac{\omega(p)}{p} V(p)\phi^{-}\left(\frac{\log(x/pd)}{\log p}\right)$$

$$+ \sum_{\substack{I < D \\ i^3 \delta(D) \geqslant x}} \sum_{p \in I} \frac{\omega(p)}{p} V(p) \phi^-\left(\frac{\log(x/pd)}{\log p}\right) .$$

The first sum in (8.6) (let $D = I_1 \ldots I_r$, $I_1 > \ldots > I_r$ as usual) is at most

$$\sum_{\substack{pd \in D \\ p < p(d)}} \frac{\omega(p)}{p} V(p) = \sum_{\substack{p \in I_r \\ p < p(d)}} \frac{\omega(p)}{p} V(p) = V\left(\frac{i_r}{z_2}\right) - V(p(d))$$

$$\leqslant V(p(d)) \left\{ \frac{\log p(d)}{\log(i_r/z_2)} \left(1 + \frac{A}{\log(i_r/z_2)}\right) - 1 \right\}$$

$$\leqslant V(p(d)) \left\{ \frac{\log z_2}{\log z_1} + \frac{A}{\log^2 z_1} \log p(d) \right\}$$

$$\ll V(p(d)) \log p(d) \frac{\log z_2}{\log^2 z_1} .$$

To estimate the second sum in (8.6) we need first to have on record that

$$(8.7) \qquad 0 < f(u_2) - f(u_1) \leqslant 2e^{\gamma} \frac{u_2 - u_1}{u_1} , \quad 0 < u_1 < u_2$$

(see "Sieve Methods" (8.2.14), or argue from (1.11) and (1.13) above). Since

$$\phi^-\left(\frac{\log(x/pd)}{\log p}\right) = 0 \quad \text{unless} \quad p^3 < x/d ,$$

an application of (8.7) with $u_1 = 2$, $u_2 = \log(x/pd)/\log p$ shows the second sum in (8.6) to be

$$\ll \sum_{\substack{I < D \\ i^3\delta(D) \geq x}} \sum_{\substack{p \in I \\ p^3 < x/d \\ p < p(d)}} \frac{\omega(p)}{p} V(p) \frac{\log(x/p^3 d)}{\log p} .$$

But $i^3\delta(D) \geq x$ implies that $x \leq p^3 d z_2^{\nu(d)+3}$, so that

$$\log(\frac{x}{p^3 d}) \leq (r + 3)\log z_2 , \quad \nu(d) = r .$$

Hence the second sum in (8.6) is

$$\ll r \frac{\log z_2}{\log z_1} \sum_{z_1 \leq p < p(d)} \frac{\omega(p)}{p} V(p) = r \frac{\log z_2}{\log z_1} (V(z_1) - V(p(d)))$$

$$\ll r \frac{\log z_2}{\log^2 z_1} V(p(d))\log p(d)$$

by (1.6), so that the entire express (8.6) is

$$\ll r \frac{\log z_2}{\log^2 z_1} V(p(d))\log p(d) , \quad \nu(d) = r .$$

It follows from (8.5) that, for $\nu \geq 1$,

$$\mathcal{E}^-_{r+1} - \mathcal{E}^-_r \ll r \frac{\log z_2}{\log^2 z_1} \sum_{\substack{d|P(z_1,z) \\ \nu(d)=r}} \frac{\omega(d)}{d} V(p(d))\log p(d) .$$

From here, arguing as in Section 5, we obtain easily that

$$\mathcal{E}^-_r \leq \mathcal{E}^-_1 + O(V(z) \frac{\log^2 z}{\log^3 z_1} \log z_2 \log(\frac{\log z}{\log z_1})) .$$

Finally, by (8.2)

$$\mathcal{E}_1^- = V(z)\phi^-\left(\frac{\log x}{\log z}\right) - V(z_1)\phi^-\left(\frac{\log x}{\log z_1}\right)$$

$$+ \sum_{I} \sum_{p \in I} \frac{\omega(p)}{p} V(p)\phi^+\left(\frac{\log x/p}{\log p}\right)$$

$$= V(z)\phi^-\left(\frac{\log x}{\log z}\right) - V(z_1)\phi^-\left(\frac{\log x}{\log z_1}\right)$$

$$+ \sum_{z_1 \leqslant p < z} \frac{\omega(p)}{p} V(p)\phi^+\left(\frac{\log x/p}{\log p}\right) \leqslant 2e^{\gamma}A\,\frac{\log z}{\log^2 z_1}\,V(z)$$

by Lemma 5.2 if $z^2 \leqslant x$. Suppose $x < z^2 \leqslant y$. Then $\phi^-\left(\frac{\log x}{\log z}\right) = 0$ and Lemma 5.2 (with $w = x^{1/2}$) implies that

$$\mathcal{E}_1 \leqslant \sum_{x^{1/2} \leqslant p < z} \frac{\omega(p)}{p} V(p)\phi^+\left(\frac{\log x/p}{\log p}\right) + 2e^{\gamma}A\,\frac{\log x^{1/2}}{\log^2 z_1}\,V(x^{1/2}).$$

Since

$$\frac{\log(x/p)}{\log p} \leqslant 1\ ,$$

we have by (1.11) and (1.6) that

$$\mathcal{E}_1 \leqslant 2e^{\gamma} \sum_{x^{1/2} \leqslant p < z} \frac{\omega(p)}{p} V(p)\,\frac{\log p}{\log(x/p)} + O\left(\frac{\log z}{\log^2 z_1}\,V(z)\right)$$

$$\leqslant 2e^{\gamma}V(z)\log z\left\{\frac{1}{\log x}\log\left(\frac{\log z}{\log(x/z)}\right) + O\left(\frac{1}{\log^2 z_1}\right)\right\}$$

by Lemma 5.1. But

$$\frac{\log z}{\log x/z} = \frac{\log z}{\log y - \log z - \log y_1} \leqslant \frac{\log z}{\log z - \log y_1} = \left(1 - \frac{\log y_1}{\log z}\right)^{-1}$$

since $y \geqslant z^2$, whence

$$\mathcal{E}_1 \ll V(z)\left(\frac{\log y_1}{\log z} + \frac{\log z}{\log^2 z_1}\right) .$$

It follows that, uniformly in r,

$$\mathcal{E}_r^- \ll V(z)\left\{\frac{\log y_1}{\log z} + \frac{\log z}{\log^2 z_1} + \frac{\log^2 z}{\log^3 z_1} \log z_2 \log\left(\frac{\log z}{\log z_1}\right)\right\}, \quad r \geqslant 1 .$$

Hence, by (8.3) and (8.2), on letting $r \to \infty$,

$$(8.8) \quad V(z_1) \sum_{D \in \mathcal{D}} \mu(D) \mathcal{X}_x^-(D) \sum_{d \in D} \frac{\omega(d)}{d}$$

$$\geqslant V(z)\left\{\phi^-\left(\frac{\log x}{\log z}\right) + O\left(\frac{\log y_1}{\log z} + \frac{\log z}{\log^2 z_1}\right.\right.$$

$$\left.\left. + \frac{\log^2 z}{\log^3 z_1} \log z_2 \log\left(\frac{\log z}{\log z_1}\right)\right)\right\} .$$

Only one more thing remains to be done. By (8.7)

$$\phi^-\left(\frac{\log x}{\log z}\right) - \phi^-\left(\frac{\log y}{\log z}\right) \geqslant -2e^\gamma \frac{\log(y/x)}{\log x} \geqslant -2e^\gamma \frac{\log y_1}{\log z}$$

so that (8.8) remains true with $\phi^-(\log y/\log z)$ in place of $\phi^-(\log x/\log z)$. Hence by (8.1),

$$S(\mathcal{A},\mathcal{P},z) \geqslant XV(z)\left\{f\left(\frac{\log y}{\log z}\right) + O\left(\frac{\log y_1}{\log z} + \frac{\log z}{\log^2 z_1}\right.\right.$$

$$\left.\left. + \left(\frac{\log z}{\log z_1}\right)^2 \left(\exp\left(-\frac{\log y_1}{\log z_1}\right) + \frac{\log z_2}{\log z_1} \log\left(\frac{\log z}{\log z_1}\right)\right)\right)\right\}$$

$$+ O\left(\frac{\log(z/z_1)}{\log z_2} 2^{\frac{\log(z/z_1)}{\log z_2}} \left| \sup_{\underline{\rho},\underline{\sigma}} \sum_{m<M} \sum_{n<N} \rho_m \sigma_n R_{mn} \right|\right)$$

The choice of y_1, z_1 and z_2 in accord with (7.8) is now a straightforward matter; indeed, we shall use precisely Motohashi's choice. Let

$$\tau = (\log\log z)^{1/10}, \tag{8.9}$$

$$z_1 = z^{\tau^{-2}}, \quad z_2 = z^{\tau^{-9}}, \quad y_1 = z^{\tau^{-1}}.$$

Then

$$S(\mathcal{A},\mathcal{P},z) \geqslant XV(z)\{f(\frac{\log y}{\log z}) + O(\frac{1}{(\log\log z)^{1/10}})\}$$

$$+ O(e^{(\log\log z)^{9/10}} \sup_{\underline{\rho},\underline{\sigma}} |\sum_{m < M} \sum_{n < N} \rho_m \sigma_n R_{mn}|) .$$

The corresponding upper bound may be derived in the same way.

Appendix

Proof of Lemma 5.1.

The sum is equal to

$$B(z_1)(V(z_1) - V(w)) + \sum_{z_1 \leqslant p < w} \frac{\omega(p)}{p} V(p) \int_{z_1}^{p} dB(t)$$

$$= B(z_1)(V(z_1) - V(w)) + \int_{z_1}^{w} (V(t) - V(w))dB(t)$$

$$= B(z_1)(V(z_1) - B(w)V(w) + \left(\int_{z_1}^{w} \frac{V(t)}{V(w)} dB(t)\right) V(w)$$

$$\leqslant V(w) \left\{ B(z_1)\frac{\log w}{\log z_1} (1 + \frac{A}{\log z_1}) - B(w) + \int_{z_1}^{w} \frac{\log w}{\log t} (1 + \frac{A}{\log t})dB(t) \right\}$$

and the integral equals

$$(1 + \frac{A}{\log w})B(w) - (1 + \frac{A}{\log z_1}) \frac{\log w}{\log z_1} B(z_1) + \int_{z_1}^{w} B(t)\log w \left(\frac{1}{t \log^2 t} + \frac{2A}{t \log^3 t}\right)dt .$$

Hence the sum is equal to

$$V(w)\log w \left\{ \int_{z_1}^{w} \left(\frac{1}{t\log^2 t} + \frac{2A}{t\log^3 t}\right)B(t)\,dt + A\frac{B(w)}{\log^2 w} \right\}$$

$$\leqslant V(w)\log w \left\{ \int_{z_1}^{w} \frac{B(t)}{t\log^2 t}\,dt + \frac{A}{\log^2 t} B(w) \right\},$$

as stated in the Lemma.

Apply to $B(p) = \phi^+(\frac{\log x/p}{\log p}) - 1, z_1 < w \leqslant x^{1/2}$, clearly a non-negative, continuous function. As p increases $\log x/\log p$ decreases and therefore $\phi^+(\frac{\log x}{\log p} - 1) - 1$ increases. The Lemma asserts that

$$\sum_{z_1 \leqslant p < w} \frac{\omega(p)}{p} V(p)\phi^+\left(\frac{\log x/p}{\log p}\right) + V(w) - V(z_1)$$

$$\leqslant V(w)\log w \left\{ \int_{z_1}^{w} \frac{\phi^+(\frac{\log x}{\log t} - 1)}{t\log^2 t}\,dt + \frac{1}{\log w} - \frac{1}{\log z_1} + \left(\phi^+\left(\frac{\log x}{\log w} - 1\right) - 1\right)\frac{A}{\log^2 z_1} \right\}.$$

Hence the first sum in Lemma 5.2 is at most

$$V(z_1) - V(w)\frac{\log w}{\log z_1} + V(w)\log w \left\{ \int_{\frac{\log t}{\log w}}^{\frac{\log x}{\log z_1}} \phi^+(u-1)\frac{du}{\log x} + \left(\phi^+\left(\frac{\log x}{\log w} - 1\right) - 1\right)\frac{A}{\log^2 z_1} \right\}$$

$$= V(w)\left\{\frac{V(z_1)}{V(w)} - \frac{\log w}{\log z_1}\right\} + V(w)\log w \left\{\phi^-(\frac{\log x}{\log z_1})\frac{1}{\log z_1}\right.$$

$$\left. - \phi^-(\frac{\log x}{\log w})\frac{1}{\log w} + (\phi^+(\frac{\log x}{\log w} - 1) - 1)\frac{A}{\log^2 z_1}\right\}$$

$$= V(z_1)\phi^-(\frac{\log x}{\log z_1}) - V(w)\phi^-(\frac{\log x}{\log w}) + V(w)\left\{\frac{V(z_1)}{V(w)} - \frac{\log w}{\log z_1}\right\}$$

$$\times \left\{1 - \phi^-(\frac{\log x}{\log z_1})\right\} + V(w)\log w(\phi^+(\frac{\log x}{\log w} - 1) - 1)\frac{A}{\log^2 z_1}$$

$$\leqslant V(z_1)\phi^-(\frac{\log x}{\log z_1}) - V(w)\phi^-(\frac{\log x}{\log w}) + \frac{V(w)\log w}{\log^2 z_1}$$

$$\times \left\{(1 - \phi^-(\frac{\log x}{\log z_1})) + (\phi^+(\frac{\log x}{\log w} - 1) - 1)\right\}$$

and the expression in parentheses is at most

$$\phi^+(\frac{\log x}{\log w} - 1) \leqslant \phi^+(1) = 2e^\gamma \quad .$$

Hence result.

Notes

1. When the remainder term is required in bilinear form (see Sections 7 and 8) this procedure is necessary in any case, so that we achieve a certain unity of method.
2. Actually, $u \gg 1$ suffices.
3. Required only for lower bound. Otherwise it suffices for u to be bounded away from 0 .
4. It is known that Theorem 6.1 is best possible in the sense that there exist sequences $\mathcal{A}^+$ and $\mathcal{A}^-$ for which the two inequalities in the Theorem hold respectively with asymptotic equality.
5. It is not true that y/d is the natural choice of y_1 . Professor Richert has pointed out in correspondence that a simpler choice avoids all the complications of Section 6 and yet leads to an exponent $\frac{1}{4}$ in Theorem 6.1 in place of $\frac{1}{10}$.

Promenade along Modular Forms and Analytic Number Theory

Henryk Iwaniec

INTRODUCTION

I was asked to address these lectures to specialists on analytic number theory who might know very little about modular forms; thus the most principal concepts and standard results need be mentioned. Under constraint of time and hot weather conditions I felt such program to be difficult for me to realize in a comprehensive form. Therefore I kept the exposition classical with lots of specializations for simplicity. The lectures consist of two parts:

- modular forms and Kloosterman sums,
- applications to problems from analytic number theory.

Concerning the first topics I shall try to give some global impressions about those methods and the results which seem to me to offer important tools for analytic number theory. In particular I wish to draw attention to sum formulae established recently by N.V. Kuznetsov [39] and R.W. Bruggeman [3]. By means of these formulae we infer mean-value theorems for Fourier coefficients of cusp forms and estimates for sums of Kloosterman sums.

The estimates of Kloosterman sums have potential applications in many diverse areas of analytic number theory. In the second part we describe a few of them:

- mean-values of the Riemann zeta-function,
- Brun-Titchmarsh theorem,
- mean-value theorems for arithmetic progressions,
- an additive divisor problem.

PART I: MODULAR FORMS AND KLOOSTERMAN SUMS

1. Holomorphic Modular Forms

Let H be the upper half plane

$$H = \{z = x + iy, \quad y > 0\} .$$

The group $PSL(2,\mathbb{R})$ acts on the Riemann sphere $C \cup \{\infty\}$ by

$$z \to \frac{az + b}{cz + d}$$

where $\begin{pmatrix} a & b \\ c & d \end{pmatrix}$ is a matrix with real coefficients and determinant 1. These maps, called linear fractional transformations, map conformally H onto itself. When H is equipped with the hyperbolic metric $dz = y^{-2}dxdy$ the linear fractional transformations become rigid motions.

Excluding the identity map $\begin{pmatrix} 1 & 0 \\ 0 & 1 \end{pmatrix}$, all others fall into one of the following classes:

elliptic: $c \neq 0$ and $|a + d| < 2$ (2 fixed points, complex conjugate),

hyperbolic: $c \neq 0$ and $|a + d| > 2$ (2 fixed points, real),

parabolic: $c = 0$ or $|a + d| = 2$ (1 fixed point, real or ∞).

In the conjugacy classes there are representatives which take on the normal forms:

elliptic $z \to K\cdot z$, $K = e^{i\theta}$ (rotation),

hyperbolic $z \to K\cdot z$, $K > 0$ (dilation),

parabolic $z \to z + b$, $b \in \mathbb{R}$ (translation).

We shall be mostly interested in the following groups:

full modular group

$$\Gamma(1) = PSL(2,\mathbb{Z}) ,$$

principal congruence group of level q

$$\Gamma(q) = \left\{ \begin{pmatrix} a & b \\ c & d \end{pmatrix} \in \Gamma(1) ; \quad \begin{pmatrix} a & b \\ c & d \end{pmatrix} \equiv \begin{pmatrix} 1 & \\ & 1 \end{pmatrix} \pmod q \right\},$$

Hecke congruence group of level q

$$\Gamma_o(q) = \left\{ \begin{pmatrix} a & b \\ c & d \end{pmatrix} \in \Gamma(1) \; ; \quad c \equiv 0 \quad (\mathrm{mod}\ q) \right\} .$$

The congruence subgroups are those subgroups of $\Gamma(1)$ containing $\Gamma(q)$ for some integer $q \geqslant 1$. They have finite index in $\Gamma(1)$; for example

$$[\Gamma(1) : \Gamma(q)] = \tfrac{1}{2} q^3 \prod_{p|q} (1 - \tfrac{1}{p^2}) \qquad \text{if} \quad q > 2 ,$$

$$[\Gamma(1) : \Gamma_o(q)] = q \prod_{p|q} (1 + \tfrac{1}{p}) , \qquad \text{any} \quad q .$$

These groups occur naturally in many arithmetical problems, for instance the Hecke groups $\Gamma_o(q)$ are very important for constructing Kloosterman sums of a type we need for problems from analytic number theory.

Now let Γ be an arbitrary subgroup of $PSL\,(2\,,\mathbb{Z})$. Two points z_1 and z_2 from $\mathbb{C} \cup \{\infty\}$ are called Γ-equivalent if

$$z_1 = \gamma z_2 \qquad \text{for some} \quad \gamma \in \Gamma .$$

Then the notion of a fundamental domain for Γ is introduced in the usual way. In the case of the modular group its fundamental domain can be chosen as

$$F = \{z = x + iy\, ; \; -\tfrac{1}{2} \leqslant x \leqslant \tfrac{1}{2} , \; |z| \geqslant 1\} .$$

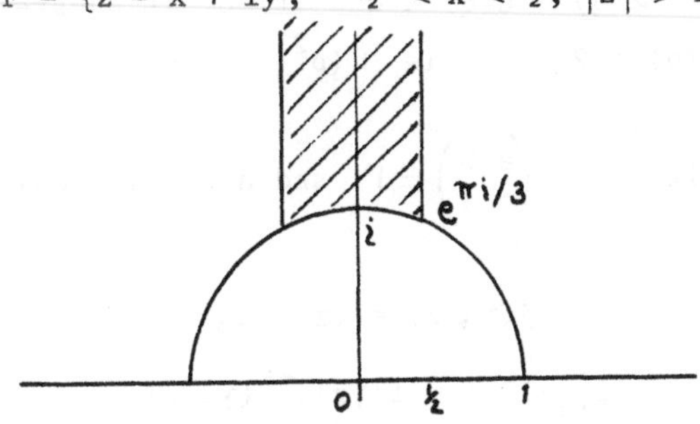

For general Fuchsian groups of the first kind (see [38]) a fundamental domain can be chosen as a polygon bounded by geodesics and having a finite (even) number of sides and special values for the interior angles. The vertices with angles zero are fixed points of parabolic transformations from Γ and are called cusps; they are placed on $\mathbb{R} \cup \{\infty\}$. The fundamental domain can be chosen so that all cusps are pairwise Γ- inequivalent.

Parabolic vertices (cusps) are very important for Fourier analysis of modular forms.

Let us give a complete set of inequivalent cusps for Hecke group $\Gamma_o(q)$ (c.f. for example [8]).

Theorem 1. Every cusp of $\Gamma_o(q)$ is equivalent to one among the following: u/w, $u,w > 0$, $w|q$, $(u, w) = 1$. Two cusps of this type u/w and u_1/w_1 are equivalent iff $w_1 = w$ and $u \equiv u_1 \mod (w, q/w)$. Hence the number of inequivalent cusps is equal to

$$\nu(q) = \sum_{w|q} \varphi((w, \tfrac{q}{w})) .$$

For later use let us denote $\mu(\mathfrak{a}) = (w, q/w)q^{-1}$ if $\mathfrak{a}$ is equivalent to u/w.

Notice that $\nu(q) = d(q) \ll q^{\varepsilon}$ if q is square-free, in particular $\nu(p) = 2$, while $\nu(p^2) = p + 1$ so it can be arbitrarily large.

For a matrix $\gamma = \begin{pmatrix} a & b \\ c & d \end{pmatrix} \in \Gamma$ and a complex number z, let us put

$$j(\gamma, z) = cz + d ,$$

$$(f|_k\gamma)(z) = f(\gamma z)j^{-k}(\gamma, z)$$

where k is a positive integer. A complex-valued function $f(z)$ is called Γ- modular form of weight k if

i) $(f|_k\gamma)(z) = f(z)$ for any $\gamma \in \Gamma$,

ii) $f(z)$ is holomorphic in H,

iii) $f(z)$ is holomorphic at each cusp;

this means that $f(z)$ is holomorphic on the Riemann surface $\Gamma\backslash H^*$. Since most investigations concern power series expansions around cusps it is convenient to change the variables appropriately. To this end send ∞ to a given cusp $\mathfrak{a}$ of Γ by means of some $\sigma_{\mathfrak{a}} \in SL(2,\mathbb{R})$ such that

$$\sigma_{\mathfrak{a}}\infty = \mathfrak{a}, \sigma_{\mathfrak{a}}^{-1}\Gamma_{\mathfrak{a}}\sigma_{\mathfrak{a}} = \left\{\begin{pmatrix}1 & n\\ 0 & 1\end{pmatrix} ; n \in \mathbb{Z}\right\} = G, \quad \text{say},$$

where $\Gamma_{\mathfrak{a}} = \{\gamma \in \Gamma;\ \gamma\mathfrak{a} = \mathfrak{a}\}$ is the stabilizer group of $\mathfrak{a}$. Such a $\sigma_{\mathfrak{a}}$ is determined up to a translation from the right. Then (iii) means that the following expansion holds:

$$(f|_k\sigma_{\mathfrak{a}})(z) = \sum_{n=0} c_{\mathfrak{a}n}\, e(nz), \quad e(z) = e^{2\pi iz}.$$

The series on the right-hand side is called Fourier series. A modular form which vanishes at each cusp $\mathfrak{a}$, i.e., with constant term

iv) $c_{\mathfrak{a}0} = 0$

is called a cusp form, so it has exponential decay at cusps.

The linear spaces $\mathcal{M}_k(\Gamma)$ and $\mathcal{M}_k^{o}(\Gamma)$ of modular forms and cusp forms respectively for a given group Γ and a given $k > 0$ are finite dimensional. In case of the modular group one must have $k \equiv 0 \pmod 2$ unless $\mathcal{M}_k(\Gamma)$ is trivial, and then

$$\dim \mathcal{M}_k(\Gamma) = \begin{cases} 1 + [k/12] & \text{if } k \not\equiv 2 \pmod{12}, \\ [k/12] & \text{if } k \equiv 2 \pmod{12}. \end{cases}$$

There are no cusp forms ($\not\equiv 0$) for $k = 2, 4, 6, 8, 10, 14$. This fact leads to interesting identities between coefficients of special forms.

The first non-trivial cusp form occurs for $k = 12$; it is

$$\Delta(z) = e(z) \prod_{n=1}^{\infty} (1 - e(nz))^{24} = \sum_{n=1} \tau(n) e(nz) .$$

The Fourier coefficients $\tau(n)$ (Ramanujan function) possess many interesting properties which attract the attention of many mathematicians.

For congruence groups (and in fact in the general case of subgroups of finite index) the dimension of the space of modular forms of weight k increases linearly as $k \to \infty$ (by the Riemann - Roch theorem, for example). This at least guarantees the existence of modular forms. The first problem which comes immediately is an explicit construction of a family of modular forms which generates the space $\mathcal{M}_k(\Gamma)$. A beautiful and simple construction goes back to Poincaré.

For any cusp $\mathfrak{a}$ of Γ, $m \geqslant 1$ and $k > 2$, the Poincaré series are defined by

$$P_{\mathfrak{a}m}(z\, ; k) = \sum_{\gamma \in \Gamma_{\mathfrak{a}} \backslash \Gamma} j(\sigma_{\mathfrak{a}}^{-1}\gamma , z)^{-k} \, e(m\sigma_{\mathfrak{a}}^{-1}\gamma z) ,$$

and the Eisenstein series are defined by

$$E_{\mathfrak{a}}(z , k) = \sum_{\gamma \in \Gamma_{\mathfrak{a}} \backslash \Gamma} j(\sigma_{\mathfrak{a}}^{-1}\gamma , z)^{-k} .$$

Let us see how these series look for the full modular group $PSL(2, \mathbb{Z})$. There is only one cusp $\mathfrak{a} = \infty$ for which, $\Gamma_\infty = G = \left\{ \begin{pmatrix} 1 & b \\ & 1 \end{pmatrix} ;\ b \in \mathbb{Z} \right\}$ and the cosets $\Gamma_\infty \backslash \Gamma$ are parametrized (one-to-one) by pairs of numbers (c , d), $c \geqslant 0$, $(d , c) = 1$ giving

$$P_m(z , k) = \sum_{c,d} (cz + d)^{-k} \, e(m\gamma z) ,$$

$$E(z , k) = \sum_{c,d} (cz + d)^{-k} \qquad \text{(essentially Epstein's zeta-function)}$$

with $\gamma = \begin{pmatrix} * & * \\ c & d \end{pmatrix} \in \Gamma$. All Poincaré series $P_m(z , k)$ are cusp forms, and the following holds:

Theorem 2. Every cusp form is a linear combination of the Poincaré series $P_m(z, k)$, $m \geq 1$.

It is very instructive to give the proof which illustrates a technique of Petersson. His greatest contribution to the theory was an introduction of an inner product on the space of cusp forms $\mathfrak{M}_k^o(\Gamma)$:

$$\langle f, g\rangle_k = \int_F f(z)\overline{g(z)}y^k dz$$

where $dz = y^{-2}dxdy$ is a $PSL(2,\mathbb{R})$ invariant measure on H, F - fundamental domain and $f, g \in \mathfrak{M}_k^o(\Gamma)$. Then $\mathfrak{M}_k^o(\Gamma)$ is a finite dimensional Hilbert space. The key point is that the product of modular form $f(z) \in \mathfrak{M}_k(\Gamma)$ against the Poincaré series $P_{\mathfrak{a}m}(z, k)$ is essentially equal to the m-th Fourier coefficient of $f(z)$ expanded at cusp $\mathfrak{a}$; precisely letting

$$(f|_k\sigma_{\mathfrak{a}})(z) = \sum_{n=0}^{\infty} c_{\mathfrak{a}n}\, e(nz)$$

we have (Petersson formula I):

$$\langle f, P_{\mathfrak{a}m}\rangle k = \frac{(k-2)!}{(4\pi m)^{k-1}} c_{\mathfrak{a}m}.$$

Now it is easy to complete the proof of Theorem 2. It follows from the following observation: any cusp form which is orthogonal to the closed space spanned by Poincaré series must have all its Fourier coefficients equal to zero; thus it vanishes identically.

There is a long history of attempts to find explicit linear relations satisfied by the Poincaré series. Unfortunately very little has been established; it is not even known when $P_m(z, k) \not\equiv 0$. For the full modular group the first Poincaré series $P_m(z, k)$ with $1 \leq m \leq \theta_k = \dim \mathfrak{M}_k^o(\Gamma(1))$ form a basis, consequently

$$P_m(z, k) \not\equiv 0 \quad \text{for} \quad 1 \leq m \leq \theta_k.$$

Rankin [50] has shown that $P_m(z, k) \not\equiv 0$ for $m < c(\varepsilon)k^{2-\varepsilon}$, $k > 14$. Actually it is conjectured (generalized Lehmer conjecture) that if $\mathfrak{M}_k^o(\Gamma)$ is not empty then all Poincaré series do not vanish identically.

Originally Lehmer formulated the conjecture for Ramanujan's τ - function $(k = 12)$ asserting that

$$\tau(n) \neq 0 \qquad \text{for any} \qquad n \geqslant 1 ,$$

which amounts to the same thing because $P_n(z ; 12) = c\tau(n)\Delta(z)$ and $\Delta(z) \neq 0$ on H.

Potential profits from Eisenstein and Poincaré series are extracted from their Fourier expansions.

There are several methods; we choose the one based on the double coset decomposition of $\sigma_{\mathfrak{a}}^{-1}\Gamma\sigma_{\mathfrak{b}}$ à la Bruhat.

Let $\mathfrak{a}$, $\mathfrak{b}$ be two cusps of Γ (not necessarily distinct) and $\Gamma_{\mathfrak{a}}$, $\Gamma_{\mathfrak{b}}$ their stabilizers in Γ. If $\mathfrak{a}$ and $\mathfrak{b}$ are Γ - equivalent put $\delta_{\mathfrak{a}\mathfrak{b}} = 1$ and if $\mathfrak{a} = \tau\mathfrak{b}$ with some $\tau \in \Gamma$ then put

$$\rho_{\mathfrak{a}\mathfrak{b}} = \sigma_{\mathfrak{a}}^{-1}\tau\sigma_{\mathfrak{b}} = \begin{pmatrix} 1 & b \\ & 1 \end{pmatrix} ,$$

otherwise put $\delta_{\mathfrak{a}\mathfrak{b}} = 0$. We than have

$$\sigma_{\mathfrak{a}}^{-1}\Gamma\sigma_{\mathfrak{b}} = \delta_{\mathfrak{a}\mathfrak{b}}\,\rho_{\mathfrak{a}\mathfrak{b}}\,G \cup \bigcup_{\gamma > 0} \; \bigcup_{\delta \,(\mathrm{mod}\ \gamma)} G\sigma G$$

where γ, δ run over numbers such that there exist α and β with

$$\begin{pmatrix} \alpha & \beta \\ \gamma & \delta \end{pmatrix} \in \sigma_{\mathfrak{a}}^{-1}\Gamma\sigma_{\mathfrak{b}} .$$

Then, for any positive γ which is left-lower corner entry of a matrix $\begin{pmatrix} * & * \\ \gamma & * \end{pmatrix}$ from $\sigma_{\mathfrak{a}}^{-1}\Gamma\sigma_{\mathfrak{b}}$, define Kloosterman sums by

$$S_{\mathfrak{a}\mathfrak{b}}(m, n ; \gamma) = \sum_{\delta \,(\mathrm{mod}\ \gamma)} e((m\alpha + n\delta)\gamma^{-1}) .$$

They are well defined and depend slightly on the choice of $\sigma_{\mathfrak{a}}, \sigma_{\mathfrak{b}}$. Equivalent cusps yield quite the same Kloosterman sums. Let the Fourier expansion of $P_{\mathfrak{a}m}(z; k)$ around the cusp $\mathfrak{b}$ be

$$j(\sigma_{\mathfrak{b}}, z)^{-k} P_{\mathfrak{a}m}(\sigma_{\mathfrak{b}} z, k) = \sum_{n=1}^{\infty} p_{\mathfrak{a}\mathfrak{b}m}(n, k) e(nz) .$$

<u>Theorem 3</u>. (<u>Petersson formula II</u>.) We have

$$p_{\mathfrak{a}\mathfrak{b}m}(n, k) = \left(\frac{n}{m}\right)^{\frac{k-1}{2}} \left\{ \delta_{mn} \delta_{\mathfrak{a}\mathfrak{b}}\, e(b_{\mathfrak{a}\mathfrak{b}} n) + 2\pi i^k \sum_{\gamma > 0}^{\Gamma} \frac{1}{\gamma} S_{\mathfrak{a}\mathfrak{b}}(m, n; \gamma) J_{k-1}\left(\frac{4\pi\sqrt{mn}}{\gamma}\right) \right\} .$$

Here $J_k(y)$ is the Bessel function of integral order;

$$J_k(y) = \frac{1}{\pi} \int_0^{\pi} \cos(y \sin\varphi - k\varphi)\, d\varphi \qquad \text{(Poisson's formula)}.$$

Notice that the series converges absolutely because

$$S_{\mathfrak{a}\mathfrak{b}}(m, n; \gamma) \ll \gamma$$

and

$$|J_k(y)| \leqslant \min\left(1, \frac{(y/2)^k}{(k-1)!}\right) .$$

The moduli γ form a well-spaced sequence so one has

$$\#\{\gamma \leqslant x\} \ll x .$$

Let us write the Kloosterman sums explicitly for Hecke groups $\Gamma = \Gamma_0(q)$. For the modular group $(\Gamma = \Gamma_0(1))$ we have one cusp equivalent to ∞ for which the Kloosterman sum $S_{\infty\infty}(m, n; c)$ is equal to the classical one

$$S(m, n; c) = \sum_{\substack{d \,(\mathrm{mod}\ c) \\ (d, c)=1}} e\left(\frac{md + n\bar{d}}{c}\right), \qquad d\bar{d} \equiv 1 \pmod{c}$$

with c being an arbitrary positive integer. For $\Gamma_0(q)$ there are more cusps, so more Kloosterman sums. The following case is quite interesting:

$$q = rs \quad \text{with} \quad (r, s) = 1, \quad \mathfrak{a} = \infty, \ \mathfrak{b} = \frac{1}{s}.$$

We then may take

$$\sigma_\infty = \begin{pmatrix} 1 & \\ & 1 \end{pmatrix}, \quad \sigma_{1/s} = \begin{pmatrix} \sqrt{r} & \\ s\sqrt{r} & 1/\sqrt{r} \end{pmatrix}$$

and after a little work on computing stabilizers and making some arithmetical rearrangements we deduce that

$$S_{\infty\, 1/s}(m, n; \gamma) = e\left(n \frac{\bar{s}}{r}\right) S(m\bar{r}, n; sc)$$

where $\gamma = s\sqrt{r}c$, c being an arbitrary positive integer. For arbitrary cusps $\mathfrak{a}$ and $\mathfrak{b}$ of $\Gamma_0(q)$ there is no simple formula for $S_{\mathfrak{a}\mathfrak{b}}(m, n; \gamma)$; it is not even easy to see the sequence of moduli γ. But if $\mathfrak{a} = \mathfrak{b}$ then γ can be any positive number from the sequence $\mu(\mathfrak{a})^{-1}\mathbb{Z}$. We recall that $\mu(\mathfrak{a}) = (w, q/w)q^{-1}$ if $\mathfrak{a} \sim u/w$, $w|q$, $(u, w) = 1$; thus γ takes integral values.

While we are on the subject of Kloosterman sums let us say a few words about upper bounds. The celebrated result of A. Weil states that

$$|S(m, n; c)| \leqslant (m, n, c)^{\frac{1}{2}}\tau(c)c^{\frac{1}{2}}.$$

It can be shown that the generalized sums $S_{\mathfrak{a}\mathfrak{a}}(m, n; \gamma)$ are expressible as the product of the classical ones with moduli $p^{\nu_p} \| \gamma$. Thus

$$|S_{\mathfrak{a}\mathfrak{a}}(m, n; \gamma)| \leqslant (m, n, \gamma)^{\frac{1}{2}}\tau(\gamma)\gamma^{\frac{1}{2}}.$$

The Petersson formula II from Theorem 3 can be used for bounding Fourier coefficients of cusp forms. By A. Weil's estimate for Kloosterman sums it easily follows that

$$p_{\mathfrak{a}\mathfrak{a}m}(n, k) \ll n^{\frac{k-1}{2} + \theta + \varepsilon}$$

with $\theta = \frac{1}{4}$, and, by Theorem 2, the same bound holds for the n-th coefficient of any cusp form (obviously the constant implied in $\ll$ may depend on the group as well as on the form itself).

The result with $\theta = \frac{3}{10}$ was proved in 1939 by Rankin [49] before A. Weil's estimate was known. Rankin's method doesn't depend on Kloosterman sums, and it works in much more general situations (we shall speak of Rankin's method later).

For a long time it has been conjectured, known as the Ramanujan - Petersson conjecture, that the n-th coefficient of any cusp form of weight $k \geq 2$, (k - integer) with respect to a congruence group is

$$a_n \ll n^{\frac{k-1}{2} + \varepsilon} .$$

Only in 1971 P. Deligne [6] first reduced the conjecture to the Riemann - Weil hypothesis for algebraic varieties, and, finally, in 1974 [7], he proved the latter. Actually a more precise bound is true for coefficients of cusp forms which are eigenfunctions of Hecke operators.

To keep the size down let us introduce Hecke operators T_n for $\Gamma = \Gamma_o(q)$ with $n \geq 1$, $(n, q) = 1$. Then the operator T_n is defined on $\mathfrak{M}_k^o(\Gamma)$ as follows:

$$T_n f(z) = n^{\frac{k-1}{2}} \sum_{ad=n} d^{-k} \sum_{b \pmod d} f\left(\frac{az + b}{d} \right) .$$

These operators act from $\mathfrak{M}_k^o(\Gamma)$ onto $\mathfrak{M}_k^o(\Gamma)$; they are bounded hermitian, multiplicative in the following sense:

$$T_m T_n = \sum_{d|(m,n)} T_{mn/d^2} \quad \text{for} \quad (mn, q) = 1 ,$$

and therefore they mutually commute. Since the space of cusp forms $\mathcal{M}_k^o(\Gamma)$ is finite dimensional it is possible to choose its orthonormal basis $f_j(z)$ consisting of eigenvectors of all T_n with $(n, q) = 1$, i.e.,

$$T_n f_j(z) = \tau_j(n) f_j(z) .$$

The eigenvalues $\tau_j(n)$ are real and satisfy the same multiplicativity relation as T_n's:

$$\tau_j(m)\tau_j(n) = \sum_{d|(m,n)} \tau_j(mn/d^2) .$$

The commutative algebra of Hecke operators is generated by T_p, p-prime; indeed

$$T_m T_n = T_{mn} \qquad \text{if} \qquad (m, n) = 1, \quad (mn, q) = 1,$$

and

$$T_{p^r} T_p = T_{p^{r+1}} + T_{p^{r-1}} \qquad \text{if} \qquad r \geqslant 1, \quad p \nmid q .$$

The latter recursion formula readily shows that for any $r \geqslant 1$, and $p \nmid q$ we have

$$T_{p^r} = U_r(\tfrac{1}{2}T_p) = \sum_{0 \leqslant 2k \leqslant r} (-1)^k \frac{(r-k)!}{k!(r-2k)!} (T_p)^{r-2k}$$

where $U_r(x)$ is the Chebyshev polynomial defined by

$$U_r(\cos\vartheta) = \sin((r+1)\vartheta)/\sin\vartheta .$$

Hecke operators have many fascinating properties. Let us mention how they act on Fourier expansions of $f(z) \in \mathcal{M}_k^o(\Gamma)$ at the cusp $\mathfrak{a} = \infty$. Suppose

$$f(z) = \sum_{m=1}^{\infty} a_m e(mz)$$

and

$$T_n f = \sum_{m=1}^{\infty} b_m e(mz) .$$

Then

$$b_m = n^{-\frac{k-1}{2}} \sum_{d|(m,n)} d^{k-1} a_{mn/d^2} .$$

In particular if $f(z)$ is an eigenvector of all T_n with eigenvalue $\lambda(n)$ then

$$\sum_{d|(m,n)} d^{k-1} a_{mn/d^2} = n^{\frac{k-1}{2}} \lambda(n) a_m .$$

Taking $m = 1$ it gives

$$a_n = a_1 \lambda(n) n^{\frac{k-1}{2}} .$$

So, after normalization $a_1 = 1$ (if possible) the coefficients a_n are multiplicative $((n , q) = 1)$. The fun begins when n is not coprime with q. The operators T_n are not normal and $\mathcal{M}_k^o(\Gamma_o(q))$ might not have a basis consisting of eigenvectors of all T_n. These cases have been worked out by Atkin and Lehner [1]. Since we do not need their theory in our exposition, we limit ourselves to what has been said.

Now we are able to give another version of the Petersson conjecture; namely, that

$$|\lambda(n)| \leqslant \tau(n) \quad \text{- divisor function.}$$

The equivalence of this with the apparently weaker statement

$$|\lambda(n)| \ll n^{\varepsilon}$$

is a nice exercise. Putting $\lambda(p) = 2 \cos \theta$, enough to show that θ is real. To this end we proceed as follows.

$$p^{\varepsilon r} \gg \lambda(p^r) = \sin((r+1)\theta) / \sin \theta \gg \exp(r \operatorname{Im}\theta) ,$$

whence $\operatorname{Im}\theta = 0$ because ε can be taken arbitrarily small.

Let us recall the Petersson formula for coefficients of Poincaré series in case of the full modular group $\Gamma = \Gamma_o(1)$; namely we have

$$p_m(n, k) = \left(\frac{n}{m}\right)^{\frac{k-1}{2}} \{\delta_{mn} + 2\pi i^k \sum_{c=1}^{\infty} \frac{1}{c} S(m, n; c) J_{k-1}\left(\frac{4\pi\sqrt{mn}}{c}\right)\} .$$

As already said, A. Weil's bound for $S(m, n; c)$ yields

$$p_m(n) \ll n^{\frac{k-1}{2} + \frac{1}{4} + \varepsilon} ,$$

while Deligne's result is sharper by $\frac{1}{4}$ in the exponent. This suggests that there must be great cancellation of terms in series of Kloosterman sums weighted by the Bessel function $J_{k-1}(x)$. In the range of large argument x the Bessel function $J_{k-1}(x)$ highly oscillates; for

$$J_{k-1}(x) \sim \sqrt{\frac{2}{\pi x}} \cos\left(x - \frac{\pi}{2} k + \frac{\pi}{4}\right) \quad \text{as} \quad x \to \infty .$$

But it is most probable that the cancellation in the series is due to variation of signs of Kloosterman sums.

In 1962 Yu.V. Linnik [43] and shortly afterwards (at least in print) A. Selberg [52] expressed the following relevant conjecture.

Conjecture. (Linnik and Selberg.) For $T \geqslant (m, n)^{\frac{1}{2}}$ we have

$$\sum_{c \leqslant T} \frac{1}{c} S(m, n; c) \ll T^{\varepsilon} .$$

The constant implied in $\ll$ depends on ε alone.

Two natural questions arise whether the conjecture implies or is implied by the Petersson - Ramanujan one. Recently Ram Murty proved that the Linnik - Selberg conjecture implies the Ramanujan - Petersson conjecture for the modular group.

That a certain cancellation in sums of Kloosterman sums does exist was in fact predicted by A. Selberg in [52], where he described a relation between Kloosterman sums and eigenvalues of the non-euclidean Laplacian asserting that, for $\sigma > \frac{1}{2}$, the Dirichlet series

$$Z(m, n, s) = \sum_{1}^{\infty} c^{-2s} S(m, n; c)$$

admits an analytic continuation to $\operatorname{Re} s > \sigma$ if and only if the least positive eigenvalue λ_1 of the Laplacian is $\geqslant \sigma(1 - \sigma)$. We shall speak of the Selberg-Kloosterman zeta function $Z(m, n, s)$ at another occasion.

The Linnik-Selberg conjecture is very important for problems from analytic number theory and it is admittedly very deep. At first glance a naïve idea comes to mind -- to try to attack the conjecture by exploiting the free parameter k in Petersson formula II. To this end one intends to create a linear combination of Bessel functions $J_{k-1}(x)$ which would very well approximate to the step function from the conjecture. Unfortunately this is impossible. An accurate analysis concerning the relevant approximations has been made by N.V. Kuznetsov [39].

Let us consider the space $L^2(\mathbb{R}^+, \frac{dx}{x})$ with logarithmic measure $x^{-1}dx$. For any complex numbers μ, ν with $\operatorname{Re}(\mu + \nu) > 0$ we have

$$\int_0^{\infty} J_\mu(x) J_\nu(x) \frac{dx}{x} = \frac{2}{\pi} \frac{\sin \frac{\pi}{2}(\nu - \mu)}{\nu^2 - \mu^2} .$$

Hence, the Bessel functions $J_{2n-1}(x)$ of positive odd order form an orthogonal system. This system is not complete, because any function of the type

$$F(x, t) = \frac{\pi i}{2 \operatorname{sh} \pi t} [J_{2it}(x) - J_{-2it}(x)]$$

is orthogonal to any $J_{2n-1}(x)$. It turns out, however, that linear combinations of $F(x, t)$ with respect to continuous parameter t yield all nice functions orthogonal to $J_{2n-1}(x)$, $n \geqslant 1$. More precisely, we have

Theorem 4 (Kuznetsov). Let $f(y)$ be of C^2 class on $(0, \infty)$ such that

$$f(0) = 0, \quad f^{(p)}(y) \ll y^{-B} \quad \text{as} \quad y \to \infty, \quad B > 2, \quad p = 0, 1, 2.$$

Put

$$\hat{f}(t) = \int_0^\infty F(y, t) f(y) \frac{dy}{y}, \quad \tilde{f}(l) = \int_0^\infty J_l(y) f(y) \frac{dy}{y}.$$

Then

$$f(x) = \sum_{\substack{l = 1 \\ l \equiv 1 \pmod 2}}^{\infty} 2l\tilde{f}(l) J_l(x) + \frac{4}{\pi^2} \int_0^\infty F(x,t)\hat{f}(t) \operatorname{th}(\pi t) t\,dt$$

$$= f_B(x) + f_H(x), \quad \text{say.}$$

Let us observe at this occasion that

$$f_B(x) = \int_0^\infty f(y) \int_0^1 \xi x J_o(\xi x) J_o(\xi y) d\xi\, dy.$$

From the above discussion it is clear that Petersson's type formulae involving sums of Kloosterman sums weighted by Bessel functions of purely imaginary order $J_{it}(x)$ with continuous parameter t are needed to approach our goal (Linnik-Selberg conjecture). To get such formulae we enter the theory of non-holomorphic modular forms.

2. Non-Holomorphic Modular Forms

We now present a brief account of Maass - Selberg theory of non-holomorphic modular forms. This theory opened a new epoch in the development of a large area of mathematics, and it uses several different mathematical languages. A very nice account of the theory can be found in a recent survey article by A.B. Venkov [56]. Here I shall try to point out most important results from spectral theory of invariant integral operators and non-euclidean Laplacian.

We first regard the upper half plane H of the complex plane $\mathbb{C}$ as the homogeneous space

$$H = PSL(2, \mathbb{R})/SO(2) ,$$

where $SO(2)$ is the maximal compact subgroup (the group of unitary matrices which fix the point i). H carries a Riemann metric ds (called the Poincaré metric) which is invariant under all actions of $PSL(2, \mathbb{R})$. It is given by

$$ds^2 = y^{-2}(dx^2 + dy^2) .$$

These actions (the linear fractional transformations) are isometries. The invariant measure connected with ds is

$$dz = y^{-2}dxdy .$$

We also introduce the geodesic distance $\rho(z, z')$, and we consider the differential operator of the metric ds (Laplace-Beltrami operator). In x, y coordinates it takes the form

$$\Delta = -y^2(\partial^2 x + \partial^2 y)$$

and the geodesic distance is given by

$$u(z, z') = \tfrac{1}{2}(\mathrm{ch}\rho(z, z') - 1) = \frac{|z - z'|^2}{yy'} .$$

Before taking up the automorphic functions let us explain some properties of Δ acting on the space of functions defined on H. Let $L^2(H, dz)$ be the space of square-integrable functions with respect to dz on H and let K be a linear operator on $L^2(H, dz)$. Then K is invariant if it commutes with all operators T_g defined by

$$T_g f(z) = f(gz), \quad g \in PSL(2, \mathbb{R}).$$

It turns out that Δ is invariant. If K is an integral operator with kernel $k(z, z')$ then K is invariant if and only if

$$k(gz, gz') = k(z, z')$$

for any $z, z' \in H$ and $g \in PSL(2, \mathbb{R})$. The remarkable result of A. Selberg [51] states that the algebra $\mathcal{K}$ of invariant operators is commutative (because of weak symmetry of H), and it is generated by one element (because H has rank 1 as symmetric space). Suppose K has the kernel $k(u(z, z'))$ (called point-pair invariant) where $k(t)$ is a nice function of a real variable t, so $K \in \mathcal{K}$. According to Selberg's results there must exist another function h such that

$$K = h(\Delta).$$

The following question is fundamental; what is the relation between k and h? Selberg gave the following answer:

<u>The Selberg - Harish - Chandra transform</u>

$$Q(w) = \int_w^\infty k(t) \frac{dt}{\sqrt{t-w}}, \qquad k(t) = -\frac{1}{\pi} \int_t^\infty \frac{dQ(w)}{\sqrt{w-t}},$$

$$g(u) = Q(e^u + e^{-u} - 2),$$

$$h(r) = \int_{-\infty}^\infty e^{\pi iru} g(u)du, \quad g(u) = \frac{1}{2\pi} \int_{-\infty}^\infty e^{-iru} h(r)dr,$$

provided $h(r)$ is analytic and it satisfies certain growth conditions; for example, any $h(r)$ such that $k(t) \in C^\infty(0, \infty)$ is admissible.

Now let us turn our attention to automorphic functions, which are related to a group acting on H. Although we need the theory for congruence groups it is nice to begin with fairly general groups Γ, the so-called Fuchsian groups of the first kind, such that

i) Γ is discrete in $PSL(2, \mathbb{R})$

ii) $\mathrm{vol}(\Gamma\backslash H) < \infty$.

The fundamental domain $F = \Gamma\backslash H$ can be compact or not (in the latter case Γ contains parabolic elements) and Γ is finitely generated. A function $f(z)$ defined on H is called <u>automorphic</u> with respect to Γ if for any $\gamma \in \Gamma$ and $z \in H$

$$f(\gamma z) = f(z) .$$

Let us pick up a fundamental domain F and consider the standard Hilbert space $L^2(F, dz)$, which can be identified with $L^2(\Gamma\backslash H)$ of automorphic functions square-integrable on F, i.e.,

$$L^2(\Gamma\backslash H) = \{f(z) ; \quad f(\gamma z) = f(z) , \quad <f, f> < \infty\}.$$

We recall that the scalar product on $L^2(\Gamma\backslash H)$ is given by

$$<f, g> = \int_F f(z)\overline{g(z)}dz .$$

Now we consider the operator $\Delta = -y^2(\partial^2 x + \partial^2 y)$ defined on the set of smooth automorphic functions $f(z)$ such that f and Δf are bounded. This set is dense in $L^2(\Gamma\backslash H)$, the operator Δ is symmetric, positive definite and it has a <u>unique</u> self-adjoint extension. Since from now on we shall be speaking only of this extension, we shall denote it by the same symbol Δ. The <u>spectral resolution</u> of Δ is fundamental for all of what follows in the Maass - Selberg theory.

A Γ - <u>modular form</u> is any automorphic function $f(z)$ on H (not necessary in $L^2(\Gamma\backslash H)$ which is an eigenfunction of the operator Δ, i.e.,

$$\Delta f = \lambda f .$$

λ will be called an eigenvalue, and $\lambda \geqslant 0$.

By spectral resolution of Δ we mean expansions of automorphic functions in terms of eigenfunctions of Δ. Nearly hopeless is the problem of explicit construction of eigenfunctions from $L^2(\Gamma \backslash H)$, and very unsatisfactory is our knowledge of eigenvalues λ. A lot of information about mean-values for λ's of different kind can be inferred from Selberg's trace formula, which we intend to present soon.

But first let us give some typical examples of modular forms. To this end observe that y^s is an eigenfunction of Δ ;

$$\Delta y^s = s(1 - s) y^s$$

for any complex s with eigenvalue $\lambda = s(1 - s)$. Let $\mathfrak{a}$ be a cusp of Γ. Define the Eisenstein - Maass series by

$$E_{\mathfrak{a}}(z , s) = \sum_{\gamma \in \Gamma_{\mathfrak{a}} \backslash \Gamma} y(\sigma_{\mathfrak{a}}^{-1} \gamma z)^s .$$

The series converges absolutely in the half-plane $\mathrm{Re}\, s > 1$ (because Γ is discrete) and it defines a modular form (because Δ is $PSL(2, \mathbb{R})$ invariant) with eigenvalue $\lambda = s(1 - s)$, i.e.,

$$\Delta E_{\mathfrak{a}}(z , s) = \lambda E_{\mathfrak{a}}(z , s) .$$

3. Spectral Decomposition of $L^2(\Gamma \backslash H)$

Much as on $L^2(H, dz)$ we introduce the commutative ring of integral operators K_Γ on $L^2(\Gamma\backslash H)$. Each one is given by the automorphic kernel

$$k_\Gamma(z , z') = \sum_{\gamma \in \Gamma} k(u(z , \gamma z')) , \quad z , z' \in F .$$

As in the case of the homogeneous space $H = PSL(2, \mathbb{R})/SO(2)$, we have

$$K_\Gamma = h(\Delta) \ .$$

A proof of spectral decomposition of $L^2(\Gamma\backslash H)$ with full details is very difficult. It depends on deep results of an analytic nature concerning Eisenstein series (the meromorphic continuation, distribution of poles and the functional equation). But it is possible to give quickly a rough decomposition which already contains the most important concept, namely that of <u>cusp form</u>.

If Γ does not have cusps ($\Gamma\backslash H$ is compact), then Δ is an unbounded operator on $L^2(\Gamma\backslash H)$ with purely discrete spectrum. The spectral theory is quite simple and of little interest for our studies, for it lacks Fourier expansions and Kloosterman sums.

Therefore, let us assume that Γ has cusps. Let $\mathfrak{a}$ be one of them. Suppose $\psi \in C^\infty(0,\infty)$; we define <u>incomplete theta</u> series by

$$\theta_{\mathfrak{a}\psi}(z) = \sum_{\gamma \in \Gamma_{\mathfrak{a}}\backslash\Gamma} \psi(y(\sigma_{\mathfrak{a}}^{-1}\gamma z)), \quad y(z) = \operatorname{Im} z ;$$

thus $\theta_{\mathfrak{a}\psi}(z) \in L^2(\Gamma\backslash H)$. Let Θ be the closed subspace of all $\theta_{\mathfrak{a}\psi}$ in $L^2(\Gamma\backslash H)$ and let $f \in L^2(\Gamma\backslash H)$ be orthogonal to Θ. Hence

$$\begin{aligned}
0 = \langle\theta_{\mathfrak{a}\psi}, f\rangle &= \int_F \theta_{\mathfrak{a}\psi}(z)\overline{f(z)}dz \\
&= \sum_{\gamma \in \Gamma_{\mathfrak{a}}\backslash\Gamma} \int_F \psi(y(\sigma_{\mathfrak{a}}^{-1}\gamma z))\overline{f(z)}dz \\
&= \sum_{\gamma \in \Gamma_{\mathfrak{a}}\backslash\Gamma} \int_{\sigma_{\mathfrak{a}}^{-1}\gamma F} \psi(y)\overline{f(\gamma^{-1}\sigma_{\mathfrak{a}}z)}dz \\
&= \int_{G/H} \psi(y)\overline{f(\sigma_{\mathfrak{a}}z)}dz = \int_0^\infty \psi(y)\left(\int_0^1 \overline{f(\sigma_{\mathfrak{a}}z)}dx\right)\frac{dy}{y^2}.
\end{aligned}$$

Since $\psi(y)$ is arbitrary it follows that

$$\int_0^1 f(\sigma_{\mathfrak{a}} z)\,dx = 0 .$$

This means that the constant term in the Fourier expansion of f at the cusp $\mathfrak{a}$ is zero; precisely the series

$$f(\sigma_{\mathfrak{a}} z) = \sum_n a_n(y) e(nx)$$

has

$$a_0(y) = \int_0^1 f(\sigma_{\mathfrak{a}} z)\,dx = 0 .$$

Such functions are called cusp forms. Denote by $L^2_{cusp}(\Gamma\backslash H)$ the space spanned by cusp forms. Then what was proved means

$$L^2(\Gamma\backslash H) = L^2_{cusp}(\Gamma\backslash H) \oplus \Theta .$$

The projections onto these two spaces commute with the integral operators K_Γ as well as with the Laplacian Δ, so we already have a rough spectral decomposition. It follows from a very general theorem of Gel'fand and Pyatetskii-Shapiro that Δ has purely discrete spectrum in $L^2_{cusp}(\Gamma\backslash H)$ (possibly empty one). Therefore the basis of $L^2_{cusp}(\Gamma\backslash H)$ can be chosen to consist of orthonormal eigencusp forms called Maass cusp forms:

$$L^2_{cusp}(\Gamma\backslash H) = \bigoplus_j u_j(z) .$$

Here

$$\langle u_j , u_j\rangle = 1 , \ \langle u_j , u_{j_1}\rangle = 0 \quad \text{for} \quad j \neq j_1$$

and

$$\Delta u_j = \lambda_j u_j \quad \text{with} \quad \lambda_j = s_j(1 - s_j) > 0$$

so $s_j = \frac{1}{2} + i\kappa_j$, $\lambda_j = \frac{1}{4} + \kappa_j^2$, κ_j is real if $\lambda_j \geq \frac{1}{4}$ and κ_j is purely imaginary if $0 < \lambda_j < \frac{1}{4}$. The eigenvalues $0 < \lambda_j < \frac{1}{4}$ will be called exceptional. They play a particular rôle in the theory, which we shall mention later. Further decomposition of the space Θ of incomplete theta series in terms of Eisenstein series is possible. Let us assume the following (this is the most difficult part of Selberg-Maass theory):

i) $E_{\mathfrak{a}}(z, s)$ -- has meromorphic continuation in the whole complex s-plane.

ii) The poles in the half-plane $\mathrm{Re}\, s \geq \frac{1}{2}$ lie only on the interval $(\frac{1}{2}, 1]$.

iii) $E_{\mathfrak{a}}(\sigma_{\mathfrak{b}} z, s)$ has Fourier series expansion

$$\delta_{\mathfrak{a}\mathfrak{b}}\, y^{s} + \varphi_{\mathfrak{a}\mathfrak{b}}(s) y^{1-s} + \sum_{n \neq 0} e_{\mathfrak{a}\mathfrak{b}n}(y, s) e(nx)$$

where

$$\varphi_{\mathfrak{a}\mathfrak{b}}(s) = \sqrt{\pi}\, \frac{\Gamma(s - \frac{1}{2})}{\Gamma(s)} \sum_{\gamma > 0} \gamma^{-2s} S_{\mathfrak{a}\mathfrak{b}}(0, 0; \gamma) .$$

iv) The so-called <u>scattering matrix</u> $\Phi(s) := (\varphi_{\mathfrak{a}\mathfrak{b}}(s))$, where $\mathfrak{a}$ and $\mathfrak{b}$ range over the system of inequivalent cusps, satisfies the following matrix functional equation

$$\Phi(s)\Phi(1 - s) = I .$$

v) The Eisenstein series $E_{\mathfrak{a}}(z, s)$ satisfy a vector column functional equation

$$[E_{\mathfrak{a}}(z, s)] = \Phi(s)[E_{\mathfrak{a}}(z, 1 - s)] .$$

Now we are in a position to decompose Θ into subspaces of point and continuous spectra of Δ. Let $L^2_{\mathrm{res}}(\Gamma\backslash H)$ denote the linear space (finite dimensional) of residues of all Eisenstein-Maass series at poles from $(\frac{1}{2}, 1]$; it is a subspace of Θ. Denote by $L^2_{\mathrm{Eis}}(\Gamma\backslash H)$ the orthogonal complement of $L^2_{\mathrm{res}}(\Gamma\backslash H)$ in Θ. We then have

<u>Theorem 5</u>. Suppose $\Gamma\backslash H$ is not compact (Γ has a cusp). We then have

(S1) $L^2(\Gamma\backslash H) = L^2_{cusp}(\Gamma\backslash H) \oplus L^2_{res}(\Gamma\ H) \oplus L^2_{Eis}(\Gamma\backslash H)$ with the orthogonal projections onto each of these subspaces commuting with Δ.

(S2) $L^2_{res}(\Gamma\backslash H)$ is finite dimensional (it contains constant functions corresponding to Δ- eigenvalue $\lambda_o = 0$) .

(S3) Δ has discrete spectrum in $L^2_{cusp}(\Gamma\backslash H) \oplus L^2_{res}(\Gamma\backslash H)$ and it has purely continuous spectrum in $L^2_{Eis}(\Gamma\backslash H)$ covering the half-line $[\frac{1}{4},\infty)$ with multiplicity equal to the number of cusps, the corresponding eigenfunctions being $E_{\mathfrak{a}}(z, \frac{1}{2} + it)$,

(S4) The following spectral decomposition holds:

$$k_\Gamma(z, z') = \sum_j h(\kappa_j)u_j(z)\overline{u_j(z')} + \frac{1}{4\pi}\sum_{\mathfrak{a}} \int_{-\infty}^{\infty} h(t)E_{\mathfrak{a}}(z, \tfrac{1}{2}+it)\overline{E_{\mathfrak{a}}(z',\tfrac{1}{2}+it)}dt ,$$

the convergence of the series and the integrals holding in the strong topology of $L^2(\Gamma\backslash H)$. Here $u_j(z)$ run over an orthonormal eigenbasis of $L^2_{cusp}(\Gamma\backslash H) \oplus L^2_{res}(\Gamma\backslash H)$ and $\lambda_j = \frac{1}{4} + \kappa_j^2$ are the corresponding eigenvalues.

For automorphic functions $f(z)$ such that $f(z)$ and $\Delta f(z)$ are bounded the following eigenfunction expansion holds:

$$f(z) = \sum_{j \geqslant 0} \langle f, u_j\rangle u_j(z) + \frac{1}{4\pi}\sum_{\mathfrak{a}} \int_{-\infty}^{\infty} \langle f, E_{\mathfrak{a}}(\cdot, \tfrac{1}{2} + it)\rangle E_{\mathfrak{a}}(z, \tfrac{1}{2} + it)dz$$

and the following Parseval identity holds for any $f, g \in L^2(\Gamma\backslash H)$, f, g bounded:

$$\langle f, g\rangle = \sum_{j \geqslant 0} \langle f, u_j\rangle \overline{\langle g, u_j\rangle} + \frac{1}{4\pi}\sum_{\mathfrak{a}} \int_{-\infty}^{\infty} \mathcal{E}_{\mathfrak{a}}(t, f)\overline{\mathcal{E}_{\mathfrak{a}}(t, g)}dt$$

with

$$\mathcal{E}_{\mathfrak{a}}(t, f) = \int_F f(z)\overline{E_{\mathfrak{a}}(z, \tfrac{1}{2} + it)}dz .$$

4. Selberg Trace Formula

The spectral decomposition of the automorphic kernel $k_\Gamma(z,z')$ of integral operator K_Γ readily leads (at least formally) to the Selberg trace formula. Let us indicate briefly the main steps. First assume, for simplicity, that Γ is a strictly hyperbolic group, i.e., it has only hyperbolic elements and the identity one, and Γ is Fuchsian of the first kind. Then

$$k_\Gamma(z, z') = \sum_j h(\kappa_j)u_j(z)\overline{u_j(z')} .$$

Hence integrating along the diagonal $z = z'$ yields

$$\int_F k_\Gamma(z, z)dz = \sum_j h(\kappa_j) ,$$

the left-hand side being interpreted as the matrix trace and the right-hand side is the spectral trace. At first glance it seems that we have lost lots of information by reducing ourselves to diagonal $z = z'$, which is true to some extent. We shall see later that the Bruggeman-Kuznetsov formula requires use of the spectral decomposition beyond the diagonal.

But the point in integrating along the diagonal is that the matrix trace can be effectively calculated in terms which have a geometrical meaning and which lead to the construction of the Selberg zeta-function.

We have

$$\int_F k_\Gamma(z, z)dz = \sum_{\gamma \in \Gamma} \int_F k(u(z, \gamma z))dz$$

$$= \sum_{\{\gamma\}_\Gamma} \sum_{\tau \in \Gamma_\gamma \backslash \Gamma} \int_F k(u(z, \tau^{-1}\gamma\tau z))dz$$

where $\{\gamma\}_\Gamma = \{\rho^{-1}\gamma\rho\ ;\ \rho\in\Gamma\}$ is the conjugacy class in Γ with representative γ and $\Gamma_\gamma = \{\sigma\ ;\ \sigma^{-1}\gamma\sigma = \gamma\}$ is the centralizer of γ in Γ. From this it follows that

$$\int_F k_\Gamma(z, z)dz = \sum_{\{\gamma\}_\Gamma} \int_{F_\gamma} k(u(z, \gamma z))dz$$

where F_γ is the fundamental domain of Γ_γ. Now the integral can be calculated quite explicitly in terms of h (the Selberg - Harish - Chandra transform of k). This is possible because F_γ has rather simple shape and γ can be given a canonical form. If γ = identity then $F_\gamma = F$, and easy calculations show that

$$\int_F k(u(z, z))dz = \frac{|F|}{4\pi} \int_{-\infty}^{\infty} r(\mathrm{th}\,\pi r)h(r)dr .$$

Let γ be hyperbolic. It is called primitive in Γ if γ is not a positive power of any other element of Γ. The same term refers to conjugacy classes. Any hyperbolic class $\{\gamma\}_\Gamma$ is represented by the canonical element γ such that

$$\gamma z = N(\gamma)z ,$$

with some number $N(\gamma) > 1$ called the norm of γ. Easy calculations show that if γ is the k-th power of a primitive element then

$$\int_{F_\gamma} k(u(z, \gamma z))dz = \frac{1}{k}\ \frac{\log N(\gamma)}{(N(\gamma))^{\frac{1}{2}} - (N(\gamma))^{-\frac{1}{2}}}\ g(\log N(\gamma)) .$$

Gathering together the above evaluations we arrive at Selberg's formula for strictly hyperbolic groups. Rather similar calculations can be done if the group contains elliptic elements. But it is more difficult to apply such arguments if there are parabolic elements in Γ, i.e., if there are cusps of Γ, because of the continuous spectrum. The terms involving Eisenstein series in the spectral decomposition of the automorphic kernel are not integrable along the diagonal $z = z'$ over the fundamental domain. For this reason one subtracts the contribution of these terms in the neighborhoods of cusps, and then applies the described arguments. Next one treats carefully the integrals of Eisenstein series in the vicinity of cusps by using the so-called Maass - Selberg relations, c.f. [38].

Let us state the final result of A. Selberg [51].

Theorem 6 (Selberg's trace formula). Let Γ be a Fuchsian group of the first kind with n -inequivalent cusps. Let $h(r)$ satisfy

i) $h(r) = h(-r)$,

ii) $h(r)$ is analytic in the strip $|\mathrm{Im} r| < \frac{1}{2} + \varepsilon$,

iii) $h(r) \ll (1 + |r|)^{-2-\varepsilon}$.

Then

$$\sum_j h(\kappa_j) - \frac{1}{2\pi}\int_{-\infty}^{\infty} h(r)\frac{\varphi'}{\varphi}(\tfrac{1}{2} + ir)dr = \frac{|F|}{2\pi}\int_{-\infty}^{\infty} r(\mathrm{th}\, r)h(r)dr$$

$$+ 2\sum_{\{P\}_\Gamma}\sum_{k=1}^{\infty}\frac{\log N(P)}{(N(P))^{k/2} - (N(P))^{-k/2}}\, g(k(\log N(P))$$

$$+ \sum_{\{R\}_\Gamma}\sum_{k=1}^{m-1}\frac{1}{m\sin(k/m)}\int_{-\infty}^{\infty} e^{-2\pi kr/m}\frac{h(r)dr}{1 + e^{-2\pi r}}$$

$$- g(0)n\log 4 + \tfrac{1}{2}h(0)(n - \mathrm{Tr}\,\Phi(\tfrac{1}{2}))$$

$$- \frac{n}{\pi}\int_{-\infty}^{\infty} h(r)\frac{\Gamma'}{\Gamma}(1 + ir)dr .$$

An important consequence of Selberg's trace formula is

Corollary 1. For $\lambda \geqslant \frac{1}{4}$ define

$$N_\Gamma(\lambda) := \#\{\lambda_j \leqslant \lambda\} .$$

Put $\lambda = \frac{1}{4} + r^2$ with $r \geqslant 0$. We then have

$$N_\Gamma(\lambda) - \frac{1}{4\pi} \int_{-r}^{r} \frac{\varphi'}{\varphi} (\tfrac{1}{2} + it)dt$$

$$= \frac{|F|}{4\pi} - \frac{n}{2\pi} \sqrt{\lambda} \log \lambda + O(\sqrt{\lambda}) ;$$

the constant implied in O may depend on the group Γ.

If there are no cusps then there is only a point spectrum so the integral over t must be suppressed and the formula above yields a very precise evaluation of $N_\Gamma(\lambda)$ itself. Yet, it is plausible that for congruence subgroups the integral over t contributes at most $O(\sqrt{\lambda} \log \lambda)$. However, in general we know very little about the determinant $\varphi(\frac{1}{2}+it)$ of the scattering matrix to the extent that we are even unable to deduce that $N_\Gamma(\lambda) \to \infty$ as $\lambda \to \infty$ or less the existence of the point spectrum and the cusp forms. The latter problem is solved only for a very special class of groups. Many results depend on the shape of the fundamental domain; see [56].

The second illuminating observation of the trace formula led Selberg to the construction of his famous zeta-function. It is defined by the following Euler product:

$$Z_\Gamma(s) = \prod_{\{P\}_\Gamma} \prod_{k=0}^{\infty} (1 - N(P)^{-s-k}), \quad \text{Re} s > 1$$

where $\{P\}_\Gamma$ runs over hyperbolic classes of Γ. On taking suitable test functions $h(r)$ in the trace formula one can read off the following properties of $Z_\Gamma(s)$:

- meromorphic continuation on the complex plane $\mathbb{C}$,
- non-trivial zeros are $s_j = \frac{1}{2} + i\kappa_j$ (Riemann hypothesis)
- a functional equation of the type

$$Z_\Gamma(1 - s) = \Psi_\Gamma(s) Z_\Gamma(s)$$

where $\Psi_\Gamma(s)$ is an explicitly determined function in terms of scattering matrix and special functions (unitary on the critical line).

Hence, using the technique of analytic number theory one can infer results about the distribution of norms NP of the primitive hyperbolic classes. The numbers NP can be viewed as "pseudo-primes;" they possess the same asymptotic distribution as the rational primes, i.e.,

$$\theta_\Gamma(x) = \sum_{NP \leqslant x} \log NP \sim x, \qquad \text{as} \quad x \to \infty.$$

A number of articles were devoted to establish formulae with a good error term. For survey of the topic see [56]. Since the Riemann hypothesis is true for $Z_\Gamma(s)$ one should expect the error term to be $O(x^{\frac{1}{2}+\varepsilon})$. Let us explain why the latter is not obvious. Consider the simplest case of the modular group. As in the theory of prime numbers we have the following explicit formula

$$\theta_\Gamma(x) = x + \sum_{|\kappa_j| \leqslant T} x^{s_j}/s_j + O(T^{-1} x \log^2 x),$$

subject to $1 \leqslant T < \sqrt{x}\,(\log x)^{-2}$. Here, the situation differs very much from the case of the Riemann zeta-function in that $Z_\Gamma(s)$ has many more zeros on the critical line ($Z_\Gamma(s)$ is of order 2), namely $\#\{s_j\,;\ s_j = \frac{1}{2} + i\kappa_j,\ |\kappa_j| \leqslant T\} \sim T^2/12$. This yields

$$\theta_\Gamma(x) = x + O(x^{\frac{1}{2}}T + T^{-1} x \log^2 x) = x + O(x^{3/4} \log x)$$

on taking the optimal value $T = x^{\frac{1}{4}} \log x$. From the above it is clear that in order to reduce the exponent 3/4 one cannot simply handle the sum over the zeros s_j in the explicit formula for $\theta_\Gamma(x)$ by summing up the terms with absolute values. A kind of equidistribution of s_j is needed to imply a considerable cancellation of terms x^{s_j}/s_j. Recently the speaker [31] established a result of such sort reducing the error term to $O(x^{35/48+\varepsilon})$. It is perhaps interesting to mention that the principal arguments are arithmetical in nature; they depend on A. Weil and D. Burgess estimates for character sums.

The most significant difference between the Riemann and the Selberg zeta-functions is that the latter does not have a natural Dirichlet series representation. This makes it impossible to inject the powerful methods of Dirichlet's polynomials (see [44]) into the theory of Selberg's zeta-function.

5. Exceptional Eigenvalues

The Selberg trace formula is valuable for inferring information on distribution of the eigenvalues in the statistical sense, but one gets into serious difficulty when the individual ones are in question. The following celebrated conjecture of Selberg fascinates for its own beauty and importance.

Conjecture (Selberg). Let Γ be a congruence group. Then the smallest positive eigenvalue λ_1 of Δ acting on $L^2(\Gamma\backslash H)$ is $\geqslant \frac{1}{4}$.

Let us give a short and elegant proof (due to W. Roelcke) of the conjecture for the modular group.

Theorem 7. For the full modular group we have $\lambda_1 > \frac{3\pi^2}{2}$.

Proof. Let $u(z)$ be a cusp form with eigenvalue λ:

$$\Delta u = \lambda u, \quad \langle u, u\rangle = 1, \quad u\text{-real}.$$

By Green's formula and the boundary periodic conditions it

follows that

$$\lambda = \int_F u \Delta u dz = \int_F (u_x^2 + u_y^2) dxdy .$$

The same holds for the shifted domain $F_1 = \tau F$, $\tau = \begin{pmatrix} 0 & 1 \\ -1 & 0 \end{pmatrix}$, therefore

$$2\lambda = \int_{F \cup F_1} (u_x^2 + u_y^2) dxdy \geqslant \int_I u_x^2 dxdy$$

where $I = \{z = x + iy ;\ -\tfrac{1}{2} \leqslant x \leqslant \tfrac{1}{2},\ \ y > y_o = \sqrt{3/2}\}$

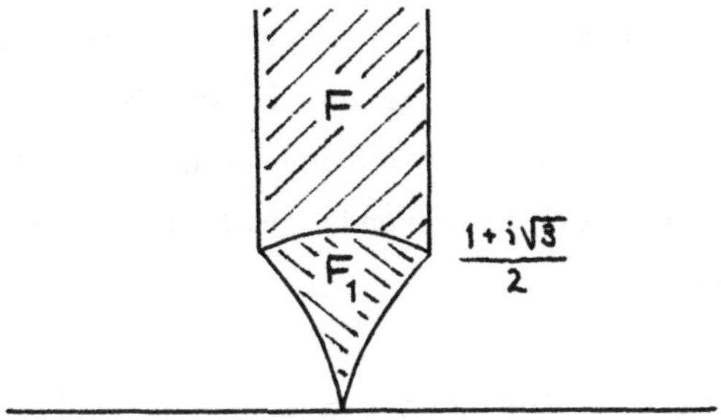

From the Fourier series expansion

$$u(x + iy) = \sum_{n \neq 0} a_n(y) e(nx)$$

it follows that

$$\begin{aligned} 2\lambda &\geqslant \int_{y_o}^{\infty} \sum_{n \neq 0} (2\pi n a_n(y))^2 dy \\ &> (2\pi y_o)^2 \int_{y_o}^{\infty} \sum_{n \neq 0} a_n^2(y) \frac{dy}{y^2} \\ &= 3\pi^2 \int_I |u|^2 dz \geqslant 3\pi^2 . \end{aligned}$$

In principle the above proof is geometrical. The peculiar shape of the fundamental domain of the modular group does matter. Recently M. Huxley (1982) used Cheeger's inequality with his own refinement of the isoperimetric inequality on the hyperbolic plane and original combinatorial arguments related to actions of Hecke operators to prove Selberg's eigenvalue conjecture for $\Gamma_o(q)$ with $1 \leqslant q \leqslant 17$. Huxley's work is a masterpiece of exploiting interrelations between various mathematical theories.

A hope for proving the conjecture is offered by ideas related to liftings of various types. Still another hope is in future developments in the theory of exponential sums in several variables.

The assumption that Γ is a congruence group is crucial. B. Randol [48] constructed a group Γ which has eigenvalues arbitrarily small and as many as one wishes. That congruence groups are very special is already evident from the following result of Selberg.

Theorem 8 (Selberg). For congruence groups we have $\lambda_1 \geqslant \frac{3}{16}$. Selberg's proof utilizes upper bounds for Kloosterman sums due to A. Weil. Later we shall give a simple variant using Kuznetsov sum formula for Hecke groups $\Gamma_o(q)$.

6. Fourier Expansion of Modular Forms

The holomorphic modular forms have an expansion in terms of exponential functions $e(nz) = e^{-2\pi ny} e(nx)$, i.e.,

$$(f|_k \sigma_{\mathfrak{a}})(z) = \sum_{n=0} c_{\mathfrak{a}n} e(nz)$$

for any cusp $\mathfrak{a}$ of Γ. Similar expansions hold for non-holomorphic modular forms, the exponential functions being replaced by the Whittaker ones:

$$\omega(nz) = \sqrt{2\pi|n|y}\, K_\nu(2\pi|n|y)e(nx) .$$

Indeed, the periodicity relation $f(\gamma z) = f(z)$ for $\gamma \in \Gamma$

implies that $f(\sigma_{\mathfrak{a}} z)$ is periodic in $x \pmod 1$; thus the ordinary Fourier series

$$f(\sigma_{\mathfrak{a}} z) = \sum_n a_n(y) e(nx)$$

exists. Then the eigenequation $\Delta f = \lambda f$ leads to an ordinary differential equation of second order for $a_n(y)$:

$$a_n''(y) = \left(\frac{\lambda}{y^2} - 4\pi^2 n^2\right) a_n(y) \ .$$

This equation has two linearly independent solutions; one has exponential growth and the other has exponential decay as $y \to \infty \, (n \neq 0)$. Only the latter is admissible if we assume that $f(z)$ has at most polynomial growth at the cusps. The second solution in question is

$$\sqrt{2\pi|n|y} \, K_\nu(2\pi|n|y) \qquad \text{for} \qquad n \neq 0$$

where $\nu = s - \frac{1}{2}$ subject to $\lambda = s(1 - s)$, and $K_\nu(y)$ is a Bessel function of the second type. It should be noted that

$$K_\nu(y) = \int_0^\infty e^{-y \operatorname{ch} \xi} \cos(\nu \xi) \, d\xi$$

and

$$\sqrt{y} \, K_\nu(y) \sim \sqrt{\frac{\pi}{2}} \, e^{-y} \quad \text{as} \quad y \to \infty .$$

For $n = 0$ we admit two independent solutions y^s and y^{1-s}. We thus obtain

$$f(\sigma_{\mathfrak{a}} z) = c_{\mathfrak{a}0} y^s + c'_{\mathfrak{a}0} y^{1-s} + \sum_{n \neq 0} c_{\mathfrak{a}n} \omega(nz)$$

for any cusp $\mathfrak{a}$ of Γ and any automorphic eigenfunction $f(z)$ with at most polynomial growth at cusps. The numbers $c_{\mathfrak{a}n}$ may depend on the cusp $\mathfrak{a}$; they are called <u>Fourier coefficients</u>.

Note the curious fact that if n is much larger than y^{-1} then $\omega(nz)$ behaves asymptotically like $e(nz)$ in the holomorphic case. Also, notice that $\omega(nz)$ are Δ-eigenfunctions (separation of variables). Recall that the cusp forms are those having zero-coefficients $c_{\mathfrak{a}0} = c'_{\mathfrak{a}0} = 0$ for any cusp $\mathfrak{a}$; thus they have exponential decay in the vicinity of each cusp.

The study of Fourier coefficients $c_{\mathfrak{a}n}$ occupies a central place in the theory. So far, even in the case of the modular group, no one has succeeded in writing a Maass cusp form (or its Fourier coefficients) in terms which are visible, somewhat like the Kloosterman sums in the expansion of holomorphic Poincaré series.

The non-holomorphic Poincaré series can be defined in much the same manner; namely

$$U_{\mathfrak{a}m}(z\,;\,s) = \sum_{\gamma\in\Gamma_{\mathfrak{a}}\backslash\Gamma} (\operatorname{Im}\sigma_{\mathfrak{a}}^{-1}\gamma z)^s e(m\sigma_{\mathfrak{a}}^{-1}\gamma z)\,, \quad m \geqslant 1 \quad \operatorname{Re}s > 1.$$

$U_{\mathfrak{a}m}(z\,;\,s)$ are not cusp forms because they are not eigenfunctions of the Laplace - Beltrami operator Δ. However $U_{\mathfrak{a}m}(z\,,\,s) \in L^2(\Gamma\backslash H)$ if $\operatorname{Re}s > 1$. Moreover the non-holomorphic Poincaré series satisfy the following analogue of Petersson formulae

<u>Petersson formula I</u>. Let $u(z) \in L^2_{cusp}(\Gamma\backslash H)$ with Fourier expansion

$$u(\sigma_{\mathfrak{a}} z) = \sqrt{y} \sum_{n\neq 0} \rho_{\mathfrak{a}}(n) K_{i\kappa}(2\pi|n|y)e(nx)$$

where $\lambda = \frac{1}{4} + \kappa^2$, $\Delta u = \lambda u$. Then for $m \geqslant 1$, $\operatorname{Re}s > 1$

$$\langle u\,,\,U_{\mathfrak{a}m}(\cdot\,,\,\bar{s})\rangle = (4\pi m)^{\frac{1}{2}-s}\pi^{\frac{1}{2}}\frac{\Gamma(s-\frac{1}{2}-i\kappa)\Gamma(s-\frac{1}{2}+i\kappa)}{\Gamma(s)}\rho_{\mathfrak{a}}(m).$$

<u>Petersson formula II</u>. Let $\mathfrak{a}, \mathfrak{b}$ be cusps of Γ, $m \geqslant 1$, $\operatorname{Re}s > 1$, and

$$U_{\mathfrak{a}m}(\sigma_{\mathfrak{b}} z ; s) = \sum_{-\infty < n < \infty} a_{\mathfrak{a}\mathfrak{b}mn}(y , s)e(nx) .$$

Then

$$a_{\mathfrak{a}\mathfrak{b}mn}(y , s) = \delta_{mn}\delta_{\mathfrak{a}\mathfrak{b}}\, e(b_{\mathfrak{a}}n)y^s \exp(-2\pi ny)$$

$$+ y^s \sum_{\gamma > 0}^{\Gamma} \gamma^{-2s} S_{\mathfrak{a}\mathfrak{b}}(m,n;\gamma)B(\gamma,m,n,y,s)$$

where

$$B(\gamma , m , n , y , s) = \int_{-\infty}^{\infty} (x^2 + y^2)^{-s} e(-nx - m/\gamma^2(x + iy))dx .$$

Very similar formulae hold for Eisenstein series. Let us write Fourier expansion of Eisenstein series:

$$E_{\mathfrak{a}}(\sigma_{\mathfrak{b}} z ; s) = \delta_{\mathfrak{a}\mathfrak{b}}\, y^s + \pi^{\frac12} \frac{\Gamma(s - \frac12)}{\Gamma(s)} \varphi_{\mathfrak{a}\mathfrak{b}o}(s)y^{1-s}$$

$$+ y^{\frac12} \sum_{n \neq 0} 2\pi^s |n|^{s-\frac12}\Gamma^{-1}(s)\varphi_{\mathfrak{a}\mathfrak{b}n}(s)K_{s-\frac12}(2\pi |n| y)e(nx)$$

where $\varphi_{\mathfrak{a}\mathfrak{b}n}(s)$ is given by the Dirichlet series

$$\varphi_{\mathfrak{a}\mathfrak{b}n}(s) = \sum_{\gamma > 0}^{\Gamma} \gamma^{-2s} S_{\mathfrak{a}\mathfrak{b}}(0 , n ; \gamma) .$$

Here the Kloosterman sum $S_{\mathfrak{a}\mathfrak{b}}(m , n ; \gamma)$ with $m = 0$ reduces to the Ramanujan one. If Γ is Hecke congruence group $\Gamma = \Gamma_o(q)$ then one can easily prove that

$$S_{\mathfrak{a}\mathfrak{a}}(0 , n ; \gamma) \ll (n , \gamma) .$$

This yields an analytic continuation of $\varphi_{\mathfrak{a}\mathfrak{b}n}(s)$ $(n \neq 0)$ in the half-place $\mathrm{Re}\, s > \frac12$. For $n = 0$ one can show that

$$\varphi_{\mathfrak{a}\mathfrak{b}o}(s) = c \cdot \frac{\zeta(2s-1)}{\zeta(2s)} \times \text{ finite Euler product;}$$

whence $\varphi_{\mathfrak{a}\mathfrak{b}o}(s)$ has only a single pole set at $s = 1$ in $\mathrm{Re}\, s > \frac12$. From this it follows that $E_{\mathfrak{a}}(z , s)$ are meromorphic in $\mathrm{Re}\, s > \frac{1}{2}$ with only a single pole at $s = 1$. Consequently, as mentioned before, the point spectrum comes from cusp forms alone and from the constant one.

7. Kuznetsov - Bruggeman Formula

Throughout this section $\Gamma = \Gamma_0(q)$. Let $k \geqslant 2$ be an even integer, $\theta_k(\Gamma) = \dim \mathcal{M}_k^0(\Gamma)$ and let $\psi_{jk}(z)$, $1 \leqslant j \leqslant \theta_k(\Gamma)$ be an orthonormal basis with the Fourier expansions at cusp $\mathfrak{a}$ given by

$$j(\sigma_{\mathfrak{a}}, z)^{-k} \psi_{jk}(\sigma_{\mathfrak{a}} z) = \sum_{m=1}^{\infty} \psi_{jk}(\mathfrak{a}, m) e(mz) .$$

Let us write the Poincaré series $P_{\mathfrak{a}m}(z; k)$ as a linear combination of $\psi_{jk}(z)$. By Petersson formula I it follows that

$$P_{\mathfrak{a}m}(z; k) = \frac{(k-2)!}{(4\pi m)^{k-1}} \sum_{1 \leqslant j \leqslant \theta_k(\Gamma)} \overline{\psi_{jk}(\mathfrak{a}, m)}\, \psi_{jk}(z) .$$

Hence by the Parseval identity

$$\langle P_{\mathfrak{a}m}, P_{\mathfrak{b}n} \rangle = \frac{((k-2)!)^2}{(16\pi^2 mn)^{k-1}} \sum_{1 \leqslant j \leqslant \theta_k(\Gamma)} \overline{\psi_{jk}(\mathfrak{a}, m)}\, \psi_{jk}(\mathfrak{b}, n).$$

But, by Petersson formula I once again, the left-hand side is $(k-2)!/(4\pi m)^{k-1} p_{\mathfrak{a}\mathfrak{b}m}(n; k)$, where $p_{\mathfrak{a}\mathfrak{b}m}(n; k)$ is the n-th Fourier coefficient of the m-th Poincaré series associated with the cusp $\mathfrak{a}$ expanded at the cusp $\mathfrak{b}$. This coefficient, by Petersson formula II, is explicitly written as a sum of Kloosterman sums (see Theorem 3). Combining both we get

$$\delta_{mn} \delta_{\mathfrak{a}\mathfrak{b}} e(bn) + 2\pi i^k \sum_{\gamma > 0}^{\Gamma} \gamma^{-1} S_{\mathfrak{a}\mathfrak{b}}(m, n; \gamma) J_{k-1}(4\pi \sqrt{mn}/\gamma)$$

$$= \frac{(k-2)!}{(4\pi \sqrt{mn})^{k-1}} \sum_{1 \leqslant j \leqslant \theta_k(\Gamma)} \overline{\psi_{jk}(\mathfrak{a}, m)}\, \psi_{jk}(\mathfrak{b}, n) .$$

This is already a sum formula. It holds for any $k \geqslant 2$, $k \equiv 0 \pmod 2$. A slight generalization is possible by forming linear combinations over k, giving

$$\sum_{\gamma > 0} \gamma^{-1} S_{\mathfrak{a}\mathfrak{b}}(m, n; \gamma) f_B(4\pi \sqrt{mn}/\gamma)$$

$$= \frac{1}{2\pi} \delta_{mn} \delta_{\mathfrak{a}\mathfrak{b}}\, e(bn) \int_0^{\infty} J_o(u) f(u)\, du$$

$$+ \frac{1}{2\pi} \sum_{k \equiv 0 \pmod 2} \sum_{1 \leqslant j \leqslant \theta_k(\Gamma)} i^k \frac{(k-1)!}{(4\pi\sqrt{mn})^{k-1}} \tilde{f}(k-1) \overline{\psi_{jk}(\mathfrak{a}, m)}\, \psi_{jk}(\mathfrak{b}, n)$$

where $f_B(x)$ and $\tilde{f}(l)$ are defined in Theorem 4.

The above formula has limited value in practice because it admits a rather small class of test functions $f_B(x)$. The complementary part of the formula involving test functions $f_H(x) = f(x) - f_B(x)$ can be deduced on a similar line by extending the above arguments to the space $L^2(\Gamma \setminus H)$ of non-holomorphic modular forms. However, we should emphasize that the details are rather sophisticated since the relevant spaces of cusp forms and Eisenstein series are infinite dimensional.

Let us state the Kuznetsov formula in its final form.

Theorem 9. Let $\Gamma = \Gamma_o(q)$. Take

$\psi_{jk}(z)$, $1 \leqslant j \leqslant \theta_k(\Gamma)$ -- an orthonormal basis of $\mathcal{M}_k^o(\Gamma)$,

$u_j(z)$, $1 \leqslant j < \infty$ -- an orthonormal basis of $L^2_{cusp}(\Gamma \setminus H)$,

$E_{\mathfrak{c}}(z, \frac{1}{2}+it)$, $t \in \mathbb{R}$ -- eigenpacket of $L^2_{eis}(\Gamma \setminus H)$,

where $\mathfrak{c}$ runs over a set of inequivalent cusps of Γ. Let $\psi_{jk}(\mathfrak{a}, m)$, $\rho_{j\mathfrak{a}}(m)$ and $\varphi_{\mathfrak{c}\mathfrak{a}\, m}(\frac{1}{2}+it)$ be the corresponding m-th Fourier coefficients of expansions around cusp $\mathfrak{a}$. For $f(x) \in C^{\infty}(0,\infty)$ put

$$\hat{f}(r) = \frac{\pi i}{2\mathrm{sh}\,\pi r} \int_0^{\infty} (J_{2ir}(x) - J_{-2ir}(x)) f(x) \frac{dx}{x} .$$

Then, for $m, n \geq 1$, and $\mathfrak{a}, \mathfrak{b}$ - cusps of $\Gamma = \Gamma_o(q)$ we have

$$\sum_{\gamma > 0}^{\Gamma} \frac{1}{\gamma} S_{\mathfrak{a}\mathfrak{b}}(m, n; \gamma) f(4\pi\sqrt{mn}/\gamma)$$

$$= \frac{1}{2\pi} \sum_{k\equiv 0 \pmod 2} \sum_{1\leqslant j\leqslant \theta_k(\Gamma)} \frac{(k-1)!\tilde{f}(k-1)}{(4\pi\sqrt{mn})^{k-1}} \overline{\psi_{jk}(\mathfrak{a}, m)}\, \psi_{jk}(\mathfrak{b}, n)$$

$$+ \sum_{j=1}^{\infty} \hat{f}(\kappa_j)(\mathrm{ch}\,\pi\kappa_j)^{-1} \overline{\rho_{j\mathfrak{a}}(m)}\, \rho_{j\mathfrak{b}}(n)$$

$$+ \frac{1}{\pi} \sum_{\mathfrak{c}} \int_{-\infty}^{\infty} \left(\frac{n}{m}\right)^{it} \overline{\varphi_{\mathfrak{c}\mathfrak{a}m}(\tfrac{1}{2}+it)}\, \varphi_{\mathfrak{c}\mathfrak{b}n}(\tfrac{1}{2}+it) \hat{f}(t)\,dt .$$

The formula of R. Bruggeman [3] is slightly weaker; it needs test functions to be holomorphic satisfying certain growth conditions. Also Bruggeman's arguments are somewhat different. Of all proofs known hitherto one which came close to Selberg's ideas from trace formula was given by D. Zagier. Let us briefly describe the initial steps of

Zagier's proof of Kuznetsov's sum formula: It starts from spectral decomposition of the automorphic kernel

$$k_{\Gamma}(z, z') = \sum_{\gamma\in\Gamma} k(u(z, \gamma z'))$$

$$= \sum_j h(\kappa_j) u_j(z) \overline{u_j(z')} + \frac{1}{4\pi} \sum_{\mathfrak{c}} \int_{-\infty}^{\infty} h(t) E_{\mathfrak{c}}(z, \tfrac{1}{2}+it) \overline{E_{\mathfrak{c}}(z', \tfrac{1}{2}+it)}\,dt .$$

When deriving Selberg's trace formula we integrated both sides along the diagonal $z = z'$, $z \in F$. Now we pick up the n-th and the m-th Fourier coefficients of both sides expanded around cusps $\mathfrak{a}$ and $\mathfrak{b}$ respectively. This means we evaluate the kernel k_Γ at points $\sigma_{\mathfrak{a}} z$, $\sigma_{\mathfrak{b}} z'$, multiply it by additive characters $e(-nx)$, $e(mx')$ and integrate in x and x' along $[0, 1]$. From the right-hand side we obtain

$$\sum_j h(\kappa_j)\overline{\rho_{j\mathfrak{a}}(m)}\,\rho_{j\mathfrak{b}}(n)K_{i\kappa_j}(2\pi\,|m|y')K_{i\kappa_j}(2\pi\,|n|y)$$

$$+\sum_{\mathfrak{c}}\int_{-\infty}^{\infty}\frac{h(t)}{|\Gamma(\frac{1}{2}+it)|^2}\left(\frac{n}{m}\right)^{it}\overline{\varphi_{\mathfrak{ca}m}(\tfrac{1}{2}+it)}\,\varphi_{\mathfrak{cb}n}(\tfrac{1}{2}+it)K_{it}(2\pi|m|y')K_{it}(2\pi|n|y)$$

and from the left-hand side we obtain

$$(yy')^{-\frac{1}{2}}\int_0^1\int_0^1 k_\Gamma(\sigma_{\mathfrak{a}} z\,,\,\sigma_{\mathfrak{b}} z')e(-nx + mx')dxdx'$$

$$=(yy')^{-\frac{1}{2}}\sum_{\gamma\in\Gamma}\int_0^1\int_0^1 k(u(\sigma_{\mathfrak{a}} z\,,\,\gamma\sigma_{\mathfrak{b}}\, z'))e(-nx + mx')dxdx'$$

$$=(yy')^{-\frac{1}{2}}\sum_{G\backslash\sigma_{\mathfrak{a}}^{-1}\Gamma\sigma_{\mathfrak{b}}/G}\int_{-\infty}^{\infty}\int_{-\infty}^{\infty}k(u(z\,,\,\tau z'))e(-nx + mx')dxdx'\ .$$

Here τ runs over representatives of the double cosets $G\backslash\sigma_{\mathfrak{a}}^{-1}\Gamma\sigma_{\mathfrak{b}}/G$. By the decomposition à la Bruhat we find that the summation is taken over $\tau = \begin{pmatrix}1 & b\\ 0 & 1\end{pmatrix}$ with some $b = b_{\mathfrak{ab}}$ if $\mathfrak{a}$ and $\mathfrak{b}$ are Γ-equivalent and over $\tau = \begin{pmatrix}* & *\\ c & d\end{pmatrix}$ with $c > 0$, $d(\mathrm{mod}\ c)$. Therefore, by $\tau z' = \frac{a}{c} - c^{-2}(x' + \frac{d}{c} + iy')^{-1}$, on shifting $x \to x + \frac{a}{c}$ and $x' \to x' - \frac{d}{c}$ we get

$$(yy')^{-\frac{1}{2}}\,\delta_{\mathfrak{ab}}e(mb_{\mathfrak{ab}})\iint_{-\infty}^{\infty}k(x + iy\,,\,x + iy')e(-nx + mx')dxdx'$$

$$+(yy')^{-\frac{1}{2}}\sum_{c>0}^{\Gamma}S_{\mathfrak{ab}}(n,m;c)\iint_{-\infty}^{\infty}k(x + iy, -c^{-2}(x'+iy')^{-1})e(-nx+mx')dxdx'.$$

Gathering the above formulae we arrive at a relation which resembles the sum formula. What is left to do is to make use of the free parameters y, y' (real valued) to transform the weights attached at Kloosterman sums into ones we wish to get. To this end integrate along the diagonal $y = \frac{n}{m}y'$ and choose

$k(t)$ appropriately. How to find $k(t)$ is not obvious, but is only a matter of pure analysis, so we stop here. D. Zagier pushed the arguments further, getting a sum formula with sufficiently explicit test functions.

Concluding this section we would like to point out the very important work of Bruggeman [4] in which a successful effort is made to give "mathematical" interpretations of various objects occurring in sum formulae; this includes a better understanding of Kuznetsov - Bessel transforms. Generalizations for other groups and automorphic models are established in [4] as well.

8. Proof of Selberg's Lower Bound $\lambda_1 \geqslant 3/16$

As an immediate application of Kuznetsov's formula we show that the smallest positive eigenvalue λ_1 for congruence groups is $\geqslant 3/16$. We take a test function $f(x) \in C^{\infty}(X, 2X)$ such that $f(x) \geqslant 0$, $f^{(\nu)}(x) \ll X^{-\nu}$, $\nu = 0, 1, 2, \dots$, where X is a sufficiently small positive parameter to be chosen later. Then for any $r > 0$ we have

$$\tilde{f}(r),\ \hat{f}(r) \ll (1 + r)^{-c} |\log X|, \quad c \text{ - any constant}$$

and for $0 < r < \frac{1}{2}$ we have

$$c_1 X^{-2r} < \hat{f}(ir) < c_2 X^{-2r} |\log X|, \quad c_1, c_2 \text{ - some positive constants.}$$

We apply Kuznetsov's formula with $m = n$ and $\mathfrak{a} = \mathfrak{b} = \infty$. On the left-hand side the A. Weil estimate for Kloosterman sums $S(m, n; \gamma)$ gives

$$\text{LHS} \ll X^{-\frac{1}{2} - \varepsilon}$$

and on the right-hand side we get

$$\text{RHS} = \sum_{0 < \lambda_j < \frac{1}{4}} \frac{\hat{f}(\kappa_j)}{\operatorname{ch}\pi\kappa_j} |\rho_j(n)|^2 + O(|\log X|) .$$

Here the error term $O(|\log X|)$ comes from trivial estimates of contributions from holomorphic cusp forms, non-holomorphic cusp forms with eigenvalues $\lambda_j \geqslant \frac{1}{4}$ and the Eisenstein series. Since $\hat{f}(\kappa_j) \geqslant c_1 X^{-2|\kappa_j|}$ it follows that $|\kappa_j| \geqslant \frac{1}{4}$, i.e., $\lambda_j \geqslant 3/16$ by letting X go to zero.

At this occasion notice that the Linnik - Selberg conjecture concerning cancellation of terms in sums of Kloosterman sums would imply

$$\text{LHS} \ll X^{-\varepsilon}$$

and consequently $\lambda_1 \geqslant \frac{1}{4}$, i.e., Selberg's eigenvalue conjecture. The opposite implication is weaker; if $\lambda_1 \geqslant \frac{1}{4}$, then the Linnik - Selberg conjecture holds for sums of Kloosterman sums weighted by functions $f(\frac{c}{T})$ where $f(x)$ is a smooth compactly supported function in $(0, \infty)$.

9. Estimates for Fourier Coefficients of Cusp Forms

First we discuss the questions for the full modular group. By manipulating with appropriate test functions in the Kuznetsov sum formula we are going to play various games. In the first direction we intend to deduce estimates for Fourier coefficients $\rho_{j\mathfrak{a}}(m)$, $\rho_{j\mathfrak{b}}(n)$ from upper bounds for Kloosterman sums $S_{\mathfrak{a}\mathfrak{b}}(m, n; \gamma)$. The first time this game was played by R. Bruggeman [3] and N.V. Kuznetsov [39] who applied their formulae to infer from A. Weil's upper bound for Kloosterman sums a mean-value theorem

$$\sum_{\kappa_j \leqslant X} \frac{|\rho_j(n)|^2}{\operatorname{ch}\pi\kappa_j} = \left(\frac{X}{\pi}\right)^2 + O_\varepsilon((\sqrt{n} + X)(nX)^\varepsilon) .$$

From this follows

$$|\rho_j(n)| \ll n^{\theta+\varepsilon} \quad \text{with} \quad \theta = \tfrac{1}{4} .$$

The best θ known hitherto is due to J.-P. Serre, namely $\theta = \frac{1}{5}$, while the analogue of the Ramanujan-Petersson conjecture states the following:

Ramanujan-Petersson conjecture. For any $\varepsilon > 0$ we have $|\rho_j(n)| \ll n^{\varepsilon}$. The constant implied in $\ll$ may depend on ε and κ_j.

Another consequence of the Bruggeman-Kuznetsov mean-value theorem is that

$$\sum_j \frac{|\rho_j(n)|^2}{\kappa_j^{2+\varepsilon} \operatorname{ch} \pi \kappa_j} < \infty .$$

This together with $N_\Gamma(\lambda) \sim \lambda/12$ suggests that the order of magnitude of $\rho_j(n)$ in κ_j should be $\exp(\frac{\pi}{2}\kappa_j)$. Therefore the following conjecture is plausible.

Conjecture. For any $\varepsilon > 0$ we have

$$|\rho_j(n)| \ll (\kappa_j n)^{\varepsilon} \exp(\tfrac{\pi}{2}\kappa_j) ,$$

and the constant implied in $\ll$ depending on ε alone.

At this occasion I should mention asymptotic formulae which were obtained by Rankin's method. The method depends on analytic properties of the zeta-function

$$\mathcal{L}_j(s) = \sum_{n=1}^{\infty} \frac{|\rho_j(n)|^2}{n^s}$$

which up to some gamma factors is equal to

$$\int_F |u_j(z)|^2 E(z, s)\, dz .$$

Hence one infers meromorphic continuation and a functional equation for $\mathcal{L}_j(s)$ from the corresponding properties of the Eisenstein series $E(z, s)$. Having got those it is easy to show by methods of analytic number theory that

$$\sum_{n \leqslant N} |\rho_j(n)|^2 = cN + O(N^{\frac{3}{5} + \varepsilon})$$

where the constant $c = \frac{6}{\pi^2} \operatorname{ch}(\pi \kappa_j)$ and the one implied in O depends at most on κ_j and ε.

Rankin's mean-value theorem shows that the Petersson conjecture holds for almost all n. For deeper results of a statistical nature see Bruggeman [3]. The upper bound for an individual coefficient $\rho_j(m)$ which the Ramanujan-Petersson conjecture asserts cannot be essentially improved but one may ask whether it is possible to deduce sharper estimates for

$$F(N) = \sum_{n \leqslant N} \overline{\rho_j(n)}\, \rho_j(n + a), \quad a \neq 0$$

due to cancellation of terms. Very interesting results were established by D. Goldfeld [20] and by A. Good [22] (actually both Goldfeld and Good deal with Fourier coefficients of holomorphic cusp forms). Inspired by a number of applications we shall pay greater attention to linear forms of the type

$$G_j(N) = \sum_{N < n \leqslant 2N} a(n) \rho_j(n) .$$

That a certain cancellation may occur when $a(n)$ is a nice smooth function can be readily seen by applying a sort of Poisson formula. Let us state Poisson's formula for Fourier coefficients of Maass cusp form

$$u(z) = \sqrt{y} \sum_{n \neq 0} \rho(n) K_\nu(2\pi |n| y) e(nx)$$

for the Hecke group, $\Gamma = \Gamma_o(q)$, where $\nu = i\kappa$.

Theorem 10. Let $f(x)$ be a smooth function with compact support in $(0,\infty)$. Define

$$f^{+}(y) = -\frac{2\pi}{\operatorname{sh}\pi\nu}\int_{0}^{\infty}[J_{2\nu}(xy) - J_{-2\nu}(xy)]\,\frac{f(x)}{x}\,dx$$

and

$$f^{-}(y) = -\frac{2\pi}{\operatorname{sh}\pi\nu}\int_{0}^{\infty}[I_{2\nu}(xy) - I_{-2\nu}(xy)]\,\frac{f(x)}{x}\,dx .$$

For any $\gamma = \begin{pmatrix} a & b \\ c & d \end{pmatrix} \in \Gamma$, $c > 0$ we have

$$c\sum_{n>0}\frac{\rho(n)}{n}\,e\left(-\frac{nd}{c}\right)f\left(\frac{4\pi\sqrt{n}}{c}\right) = \sum_{n>0}\rho(n)e\left(\frac{na}{c}\right)f^{+}(\sqrt{n}) + \sum_{n>0}\rho(-n)e\left(-\frac{na}{c}\right)f^{-}(\sqrt{n}) .$$

The linear forms $G_j(N)$ that often occur in practice involve coefficients $a(n)$ which are not smooth; therefore Poisson's formula is by no means applicable. There is, however, a possibility for terms $a(n)\rho_j(n)$ to cancel when another averaging over the spectrum λ_j, $j \geq 1$ is performed, no matter what $a(n)$ are. First results were established by the speaker [34],[35].

Theorem 11. Let $\rho_j(n)$ be Fourier coefficients of orthonormal basis $u_j(z)$ of cusp forms for the modular group. For any complex numbers a_n and any $\varepsilon > 0$ we have

$$\sum_{\kappa_j \leq K}(\operatorname{ch}\pi\kappa_j)^{-1}\Big|\sum_{N<n\leq 2N}a_n\rho_j(n)\Big|^2 \ll (K^2 + N^{1+\varepsilon})\sum_{N<n\leq 2N}|a_n|^2 .$$

In view of the conjecture that $\rho_j(n)$ has the order of magnitude $O((\kappa_j n)^{\varepsilon}\exp(\frac{\pi}{2}\kappa_j))$ Theorem 11 shows that there is a cancellation between terms of the linear forms $G_j(n)$ to the effect that

$$G_j(N) \ll (KN)^{\varepsilon}(\Sigma\,|a_n|^2)^{\frac{1}{2}}$$

for almost all j with $\kappa_j \leqslant K$, provided $K > \sqrt{N}$. This result resembles the large sieve inequalities for Dirichlet's polynomials with multiplicative characters $\chi(\mathrm{mod}\ q)$; namely

$$\sum_{q \leqslant Q} \ \sum^{*}_{\chi(\mathrm{mod}\ q)} \Big| \sum_{n \leqslant N} a_n \chi(n) \Big|^2 \leqslant (Q^2 + N) \sum_{n \leqslant N} |a_n|^2 ,$$

as well as the mean-value theorem for Dirichlet's polynomials with "characters" n^{it}; namely

$$\int_0^T \Big| \sum_n a_n n^{it} \Big|^2 dt = (T + O(N)) \sum_n |a_n|^2 .$$

For this reason one may think of Fourier coefficients of cusp forms as characters, especially when the forms are eigen-functions of all Hecke operators T_n, $n \geqslant 1$, because then the coefficients $\rho_j(n)$ are proportional to the eigenvalues $\lambda_j(n)$ of T_n which are multiplicative. There is a lively area of the theory of Dirichlet's polynomial $\Sigma\, a_n \chi(n)$, $\Sigma\, a_n n^{it}$ which concerns frequencies of large values (Montgomery, Huxley, Jutila). The theory, perhaps, will inspire further studies of $\Sigma\, a_n \rho_j(n)$.

To probe the still mysterious nature of the Fourier coefficients of cusp forms we established the following hybrid large sieve (see [34]).

<u>Theorem 12</u>. Let $M \geqslant 1$, and $K \geqslant N \geqslant 1$. Then, for any complex numbers a_m, b_n we have

$$\sum_{K < \kappa_j \leqslant 2K} \frac{1}{\operatorname{ch} \pi \kappa_j} \Big| \sum_{m \leqslant M} a_m m^{i\kappa_j} \Big|^2 \Big| \sum_{n \leqslant N} b_n \rho_j(n) \Big|^2$$

$$\ll K(K + M)\Big(\sum_m |a_m|^2\Big)\Big(\sum_n |b_n|^2\Big) .$$

To see the point in the above result one needs some explanation. For any subsequence $\mathcal{K}$ of well-spaced points κ_j from $[K, 2K]$ by mean-value theorem for Dirichlet polynomials we get

$$\sum_{\kappa_j \in \mathcal{K}} \Big| \sum_{m \leqslant M} a_m m^{i\kappa_j} \Big|^2 \ll (K+M) \sum_{m \leqslant M} |a_m|^2 .$$

Since $\#\{j ; |\kappa_j| \leqslant K\} = \frac{1}{12} K^2 + O(K)$, from this follows

$$\sum_{K < \kappa_j \leqslant 2K} \Big| \sum_{m \leqslant M} a_m m^{i\kappa_j} \Big|^2 \ll D(K, M) \sum_{m \leqslant M} |a_m|^2$$

with the same $D(K, M) = K(K+M)$ as in the hybrid large sieve inequality. We interpret this fact by saying that the characters $\lambda_j(n) = \rho_j(n)/\rho_j(1)$ are "almost orthogonal" to the characters $m^{i\kappa_j}$. It is interesting to realize that the Fourier coefficients of Eisenstein series $E(z, \frac{1}{2} + it)$ (eigenfunctions of continuous spectrum) are proportional to $\lambda_{it}(n) = \sum_{n_1 n_2 = n} (n_1/n_2)^{it}$; they do not differ much from the characters n^{it}. In this case an analogue of the large sieve inequality becomes an easy consequence of the mean-value theorem for polynomials $\sum a_n n^{it}$.

We do not have time to speak about the proofs in detail. To give a rough impression I would like to mention that they make use of the specific structure of Kloosterman sums $S(m, n; c) = \sum^*_{d \,(\mathrm{mod}\ c)} e(m\bar{d}/c + nd/c)$. The argument is somewhat deeper than the large sieve inequality for the Farey points $\bar{d}/c$ and d/c.

A lot of generalizations for Fuchsian groups are possible but really interesting and most important, as it will soon become apparent, are those for Hecke congruence groups

$\Gamma = \Gamma_0(q)$. Here, there is a new aspect of the subject -- explicit dependence on q. Moreover, the exceptional spectrum (if it occurs) needs special care. The methods and results described below were established jointly with Professor J.-M. Deshouillers in [8].

Let $\Gamma = \Gamma_0(q)$, $q \geqslant 1$. Consider an orthonormal basis $\{u_j(z);\ j \geqslant 1\}$ of Masss cusp forms in $L^2(\Gamma\backslash H)$ and an orthonormal basis $\{\psi_{jk}(z);\ 1 \leqslant j \leqslant \dim \mathfrak{M}_k^0(\Gamma)\}$, of holomorphic cusp forms of weight k. Let $\mathfrak{a}$ be a cusp of Γ and

$$u_j(\sigma_{\mathfrak{a}} z) = \sqrt{y} \sum_{n \neq 0} \rho_{j\mathfrak{a}}(n) K_{i\kappa_j}(2\pi |n| y) e(nx)$$

$$\psi_{jk}(\sigma_{\mathfrak{a}} z) = \sum_{n=1}^{\infty} \psi_{jk}(\mathfrak{a}, n) e(nz)$$

be the Fourier expansions at $\mathfrak{a}$. We state, but have not been able to prove, the following

Conjecture. For any $\varepsilon > 0$ it holds

$$|\rho_{j\infty}(n)| \ll (\lambda_j nq)^{\varepsilon} q^{-\frac{1}{2}} \exp(\tfrac{\pi}{2}\kappa_j)$$

and

$$|\psi_{jk}(\infty, n)| \ll (knq)^{\varepsilon} (q(k-1)!)^{-\frac{1}{2}} (4\pi n)^{\frac{k-1}{2}}$$

the constants implied in $\ll$ depend on ε alone.

The large sieve inequalities below support the conjecture and at the same time they show that there is a cancellation of terms in linear forms with $\rho_{j\mathfrak{a}}(n)$ and $\psi_{jk}(\mathfrak{a}, n)$.

Theorem 13. (Deshouillers and Iwaniec.) For any $a_n \in \mathbb{C}$ we have

$$\sum_{k\equiv 0 \pmod 2} e^{-\frac{k-1}{K}} \frac{(k-1)!}{(4\pi)^{k-1}} \sum_{1\leq j\leq \dim \mathcal{M}_k^o(\Gamma)} \Big| \sum_{N<n\leq 2N} a_n n^{-\frac{k-1}{2}} \psi_{jk}(\mathfrak{a},n)\Big|^2$$

$$= (\tfrac{1}{2}K^2 + O(1 + \mu(\mathfrak{a})N^{1+\varepsilon})) \sum |a_n|^2$$

and

$$\sum_{|\kappa_j|\leq K} \frac{1}{\operatorname{ch}\pi\kappa_j} \Big| \sum_{N<n\leq 2N} a_n \rho_{j\mathfrak{a}}(n)\Big|^2 \ll (K^2 + \mu(\mathfrak{a})N^{1+\varepsilon}) \sum |a_n|^2 ,$$

where the constants implied in the symbols O and $\ll$ depend on ε alone. We recall that $\mu(\mathfrak{a}) = \frac{(q, q/w)}{q}$ if $\mathfrak{a}$ is equivalent to the cusp $\frac{u}{w}$, with $w|q$, $(u, w) = 1$.

A similar inequality can be formulated for Fourier coefficients of Eisenstein series; see [8].

Strikingly enough, small eigenvalues $\lambda_j = \frac{1}{4} + \kappa_j^2 \ll 1$ release cancellation of terms in $\sum a_n \rho_{j\mathfrak{a}}(n)$. This is in great contrast to the case of the modular group. The point is that there are lots of eigenvalues $\ll 1$, presumedly of order of magnitude

$$\nu(q) = [\Gamma_o(1) : \Gamma_o(q)] = q \prod_{p|q} (1 + \frac{1}{p}) ,$$

for the Weyl- Selberg formula yields

$$\#\{j ; |\kappa_j| \leq K\} \sim \frac{\nu(q)}{12} K^2 \quad \text{as} \quad K \to \infty .$$

In the next section we shall go on to illustrate a use of the large sieve for bounding sums of Kloosterman sums. In the circumstances when the sums are counted with nice smooth weights the small eigenvalues λ_j are the only significant ones. Yet, the large sieve, being far from trivial, has a defect of not distinguishing between the imaginary κ_j's which correspond to exceptional eigenvalues $\lambda_j < \frac{1}{4}$ and the real ones. The exceptional eigenvalues create exceptional main terms which import troubles if their number is beyond our control. It prompted us that such main terms can be turned to our profit. We went on to introduce weights in order to improve the large sieve inequalities for exceptional eigenvalues. A typical result is found in [8].

Theorem 14. (Deshouillers and Iwaniec.) Let $1 \leqslant N \leqslant q$ and a_n be complex numbers. We then have

$$\sum_{0 < 4i\kappa_j < 1} \left(\frac{q}{N}\right)^{2i\kappa_j} \Big| \sum_{n \leqslant N} a_n \rho_{j\infty}(n) \Big|^2 \ll q^{\varepsilon} \sum_{n \leqslant N} |a_n|^2$$

for any $\varepsilon > 0$, the constant implied in $\ll$ depending on ε alone.

When an averaging over the level q is performed we can show a still sharper inequality.

Theorem 15. (Deshouillers and Iwaniec.) Let $N, Q, X \geqslant 1$ and a_n be complex numbers. We then have

$$\sum_{q \leqslant Q} \sum^{(q)}_{0 < 4i\kappa_j \leqslant 1} X^{4i\kappa_j} \Big| \sum_{n\leqslant N} a_n \rho_{j\infty}(n) \Big|^2 \ll (NQ) \; (Q+NX) \sum_{n\leqslant N} |a_n|^2.$$

Theorem 15 shows a certain scarcity of κ_j's; this can be expressed by a kind of density theorem

Theorem 16. Let $Q \geqslant 1$, $n \geqslant 1$ and $0 < \sigma < 1$. We then have

$$\sum_{q \leqslant Q} \sum^{(q)}_{\sigma < 4i\kappa_j \leqslant 1} |\rho_{j\infty}(n)|^2 \ll Q^{1-\sigma+\varepsilon}.$$

The constant implied in $\ll$ may depend on ε and n only.

In view of this result the following conjecture seems to be plausible.

Density Conjecture. Let $N(\Gamma, \mathfrak{S})$ be the number of exceptional eigenvalues $\lambda_j = \frac{1}{4} + \kappa_j^2$ with $\mathfrak{S} < 4i\kappa_j \leqslant 1$. Then, for $\Gamma = \Gamma_o(q)$ and any $\varepsilon > 0$ we have

$$N(\Gamma_o(q), \mathfrak{S}) \ll q^{1 - \mathfrak{S} + \varepsilon} .$$

The best we have been able to prove so far is

Theorem 17. (H. Iwaniec and J. Szmidt.) For any $\varepsilon > 0$ we have

$$\sum_{0 < 4i\kappa_j \leqslant 1} q^{Ai\kappa_j} \ll q^{1+\varepsilon}$$

with $A = 24/11$, the constant implied in $\ll$ depending on ε alone (recently the speaker proved this with $A = 12/5$).

This result depends on A. Weil's estimates of exponential sums over finite fields and the famous inequality of D. Burgess [5] for character sums, c.f. [37]. Weil's estimate for Kloosterman sums alone would yield the constant $A = 2$ which was later obtained by M. Huxley (oral communication) by a different method (an application of Selberg's trace formula).

Of particular interest are large sieve inequalities for linear forms with coefficients $a_n \equiv 1$, $n \leqslant N$. Such forms occur when incomplete Kloosterman sums are concerned, and then N is about equal to the ratio of the length of the complete sum to that of the incomplete one. In this case Theorem 15 can be improved to the effect that the number $Q + NX$ can be replaced by $Q + \sqrt{N}X$. The following estimate, also interesting for its own sake, is crucial:

$$\sum_{c \leqslant C} \Big| \sum_{m \leqslant M} \sum_{n \leqslant N} S(m, n; c)\zeta^m \xi^n \Big| \ll (C + MN)C^{1+\varepsilon} ,$$

subject to $|\zeta| = |\xi| = 1$, the constant implied in $\ll$ depending on ε alone. Unexpectedly the proof of this makes use of the non-existence of exceptional eigenvalues for the modular group.

10. Kloosterman Sums

In this section we are going to use the sum formula of Kuznetsov and the estimates for Fourier coefficients of cusp forms for bounding sums of Kloosterman sums $S_{\mathfrak{a}\mathfrak{b}}(m, n; \gamma)$ associated with cusps $\mathfrak{a}, \mathfrak{b}$ of Hecke congruence group $\Gamma = \Gamma_o(q)$. When an averaging over coefficients m, n is performed we apply the large sieve inequalities to get an extra saving effect.

Motivated by a number of applications we are led to study sums of the type

$$S_{\mathfrak{a}\mathfrak{b}}(a_M, b_N; g)$$

$$= \sum^{\Gamma} \frac{1}{\gamma} \sum_{M < m \leqslant 2M} a_m \sum_{N < n \leqslant 2N} b_n g(m, n, \gamma) S_{\mathfrak{a}\mathfrak{b}}(m, n; \gamma)$$

where $g(m, n; \gamma)$ is a weight function to be specified suitably. Let us first formulate a result with $g(m, n; \gamma)$ of the type $g(m, n; \gamma) = f(4\pi\sqrt{mn}/\gamma)$ where $f(x)$ is of C^2 class such that for some $X, Y \geqslant 1$,

$$\begin{aligned} &\text{Supp } f(x) \subset [X^{-1}, 8X^{-1}] , \\ &|f(x)| \leqslant 1 , \\ &\int |f'(x)| dx \leqslant 2 , \\ &\int |f''(x)| dx \leqslant 6XY . \end{aligned}$$

All other weights $g(m, n; \gamma)$ in practice can be transformed into $f(4\pi\sqrt{mn}/\gamma)$ by any method of separation of variables.

<u>Theorem 18</u>. (Deshouillers and Iwaniec.) For sequences $\underline{\underline{a}}$ and $\underline{\underline{b}}$ of complex numbers and $g(m, n; \gamma)$ as above we have

$$|S_{\mathfrak{a}\mathfrak{b}}(\underline{\underline{a}}_M, \underline{\underline{b}}_N; g)|^2 \ll \{Y + X^{\theta}(1+\mu(\mathfrak{a})M^{1+\varepsilon})(1+\mu(\mathfrak{b})N^{1+\varepsilon})\}$$

$$\Big(\sum_{M < m \leqslant 2M} |a_m|^2\Big)\Big(\sum_{N < n \leqslant 2N} |b_n|^2\Big) .$$

Here $\theta = 4i\kappa_1$ if the smallest positive eigenvalue for $\Gamma_o(q)$ is $\lambda_1 = \frac{1}{4} + \kappa_1^2 < \frac{1}{4}$ with $0 < 4i\kappa_1 < 1$ and $\theta = 0$ otherwise.

Let us specialize this result to particular contexts. First let us consider the modular group $\Gamma = \Gamma_o(1)$ in which case exceptional eigenvalues do not occur, so $\theta = 0$. We then deduce

Corollary 1. For $M, N, T \geqslant 1$, $\varepsilon > 0$ and any complex numbers a_m, b_n we have

$$\sum_{M < m \leqslant 2M} a_m \sum_{N < n \leqslant 2N} b_n \sum_{c < \sqrt{\frac{mn}{MN}}\, T} \frac{1}{c} S(m, n; c)$$

$$\ll T^{\varepsilon}\big((MN)^{\frac{1}{2}} + (MNT)^{\frac{1}{6}}\big)\big(\Sigma |a_m|^2\big)^{\frac{1}{2}}\big(\Sigma |b_n|^2\big)^{\frac{1}{2}} ;$$

the constant implied in $\ll$ depending on ε alone.

Our result contains (apart from the factor T^{ε}) the celebrated estimate of Kuznetsov [39]

$$\sum_{c \leqslant T} \frac{1}{c} S(m, n; c) \ll_{m,n} T^{\frac{1}{6}} (\log T)^{\frac{1}{3}} .$$

The latter has been proved by another method (independent of Kuznetsov's formula) by D. Goldfeld and P. Sarnak [21].

I should mention in this connection an upper bound

$$T^{\varepsilon}\big((MN)^{\frac{1}{2}} + T\big)\big(\Sigma |a_m|^2\big)^{\frac{1}{2}}\big(\Sigma |b_n|^2\big)^{\frac{1}{2}}$$

which follows by the large sieve inequality for Farey points d/c and $\bar{d}/c$; this bound is as strong as Corollary 1 only if $MN > T^2$.

The second interesting context is that of Kloosterman sums $S(m\bar{r}, n; sc)$ with $(r, s) = 1$. As shown in §1 they are associated with cusps $\mathfrak{a} = \infty$ and $\mathfrak{b} = 1/s$ of the group $\Gamma = \Gamma_o(rs)$. By Theorem 18 we infer

Corollary 2. Let $g(m, n, c)$ be of C^6 class supported in $[M, 2M] \times [N, 2N] \times [C, 2C]$ with partial derivatives $g^{(j)}(m, n, c) \ll M^{-j_1} N^{-j_2} C^{-j_3}$ for $(j) = (j_1, j_2, j_3)$, $0 \leqslant j_1, j_2, j_3 \leqslant 2$. Suppose the smallest positive eigenvalue of $\Gamma_o(rs)$ is $\lambda_1 \geqslant \frac{1}{4}$. We then have

$$\sum_m \sum_n \sum_{(c,r)=1} a_m b_n g(m, n, c) S(m\bar{r}, n; sc)$$

$$\ll C^{1+\varepsilon}(s\sqrt{r} + \sqrt{s(M+N)} + \sqrt{MN})(\Sigma |a_m|^2)^{\frac{1}{2}}(\Sigma |b_n|^2)^{\frac{1}{2}},$$

the constant implied in $\ll$ depending on ε alone.

If $\lambda_1 < \frac{1}{4}$ then the above upper bound is weakened by the factor $(1 + Cs\sqrt{r}/\sqrt{MN})^{2i\kappa_1}$. This loss can be reduced substantially by means of the weighted large sieve inequality provided an averaging over r and s is performed. Yet, a greater saving is obtained for incomplete Kloosterman sums

$$\sum_{(d,sc)=1} g(d) e(m \frac{\overline{rd}}{sc})$$

where $g(d)$ is a nice smooth function supported in $[D, 2D]$ with $D < sC$. The incomplete sum can be complete by any standard method, for instance by applying Poisson's formula for arithmetic progressions. The resulting sums are essentially the following

$$N^{-1} \sum_{|n| \leqslant N} S(m\bar{r}, n; sc)$$

where $N = sc/D$. Therefore Theorem 15 is applicable in its strongest version giving

Theorem 19 (Deshouillers and Iwaniec). Let $C, D, M, R, S \geqslant 1$, and $g(c, d)$ be a smooth function supported in $[C,2C] \times [D,2D]$ with derivatives $|g^{(j)}(c, d)| \ll C^{-j_1} D^{-j_2}$. We then have

$$\sum_{\substack{R < r \leqslant 2R \\ S < s \leqslant 2S \\ (r,s) = 1}} \sum_{m \leqslant M} b_{mrs} \sum_{(rd,sc) = 1} g(c, d) e(m \frac{\overline{rd}}{sc})$$

$$\ll (CDMRS)^{\varepsilon} K(C, D, M, R, S) (\sum_{r,s,m} |b_{mrs}|^2)^{\frac{1}{2}}$$

where $K^2(C, D, M, R, S) = CS(RS + M)(C + DR) + D^2 MRS^{-1} + C^2 DS\sqrt{R(RS + M)}$, the constant implied in $\ll$ depending on ε alone. If the Selberg conjecture $\lambda_1 \geqslant \frac{1}{4}$ is true then the last term $C^2 DS\sqrt{R(RS + M)}$ can be skipped.

In applications of the large sieve inequalities for Fourier coefficients of cusp forms of $\Gamma = \Gamma_0(q)$ to estimate Kloosterman sums we only used their power in the q-aspect; the K-aspect would be important as well if sums of Kloosterman sums with highly oscillating weights were considered. Such cases need more exploration.

Before leaving Kloosterman sums, I would like to point out the celebrated conjecture of C. Hooley [30] about the order of magnitude of incomplete Kloosterman sums.

Conjecture R^* (C. Hooley). Let $m \neq 0$ and $1 \leqslant D < c$. We then have

$$\sum_{\substack{1 \leqslant d \leqslant D \\ (d,c)=1}} e(m\bar{d}/c) \ll (m\,,\,c)^{\frac{1}{2}} D^{\frac{1}{2}} c^{\varepsilon}\ .$$

The conjecture presents a great contrast to the behavior of incomplete Weyl's sums

$$\sum_{1 \leqslant d < D} e\left(\frac{f(d)}{c}\right)$$

where $f(x)$ is a polynomial. In the latter sum one needs large D for $f(d)$ to grow above c in order to produce a cancellation of terms while in the Kloosterman sums $\bar{d}$ jumps over residue classes $(\bmod\ c)$ uniformly from the very beginning.

Our Theorem 19 contains an estimate for sums of incomplete Kloosterman sums which are not obtainable from Hooley's conjecture.

An inspection of various heuristical arguments led us to formulate the following desired conjecture.

Conjecture. Let $m \geqslant 1$, $r\,,\,s \geqslant 1$, $(r\,,\,s) = 1$, $1 \leqslant D \leqslant C$, and t be any real number. For any $\varepsilon > 0$ we have

$$\sum_{\substack{1 \leqslant c \leqslant C \\ (c,r)=1}} c^{it} \sum_{\substack{1 \leqslant d \leqslant D \\ (rd,sc)=1}} e\left(m\,\frac{\overline{rd}}{sc}\right) \ll (mrs(|t|+1))^{\varepsilon} C^{1+\varepsilon}\ ,$$

the constant implied in $\ll$ depending on ε alone.

PART II: APPLICATIONS

11. Power Mean-Values of the Riemann Zeta-Function

We shall discuss here some recent applications of the inequalities for Fourier coefficients of cusp forms and sums of Kloosterman sums to results on the average size of the Riemann zeta-function $\zeta(\frac{1}{2} + it)$ and Dirichlet's polynomials

$$M(s) = \sum_{M < m \leq 2M} a_m m^{-s} .$$

We shall demonstrate a few details of the arguments of Deshouillers and Iwaniec (see [10], [11]) used for bounding the integral

$$S(T, M) = \int_T^{2T} |\zeta(\tfrac{1}{2} + it)|^4 |M(it)|^2 dt .$$

The zeta-function can be well approximated by its partial sums of the type $L(it)L^{-\frac{1}{2}}$ where

$$L(it) := \sum_l b(l) l^{it}$$

and $b(l)$ is a smooth function supported in $[L, 2L]$, $L \leq T^{\frac{1}{2}}$, with derivatives $b^{(j)}(l) \ll l^{-j}$, $j = 0, 1, 2, \ldots$. Therefore the problem reduces to estimating the integrals

$$S(T, L, M) = \int f(t) |L(it)|^4 |M(it)|^2 dt$$

where the kernel function $f(t)$ is smooth, supported in $[\frac{1}{2}T, 3T]$. Squaring out one gets terms

$$b(l_1)b(l_2)b(l_3)b(l_4) a_{m_1} \bar{a}_{m_2} \left(\frac{l_1 l_2 m_1}{l_3 l_4 m_2} \right)^{it}$$

to be integrated over t with weight $f(t)$. The diagonal terms, i.e., such that $l_1 l_2 m_1 = l_3 l_4 m_2$, contribute to the main term. Those causing oscillations, i.e., such that $l_1 l_2 m_1 = l_3 l_4 m_2 + h$ with $|h| \geqslant H := L^2 M T^{\varepsilon - 1}$, contribute very little after integration over t. The remaining terms satisfy

$$l_1 l_2 m_1 = l_3 l_4 m_2 + h , \qquad 0 < |h| < H .$$

They require a delicate treatment. For simplicity we use the following elementary approximations

$$\left(\frac{l_1 l_2 m_1}{l_3 l_4 m_2}\right)^{it} = e^{ith/l_3 l_4 m_2} + O(T^{\varepsilon - 1})$$

and

$$b(l_1) = b\left(l_3 \frac{l_4 m_2}{l_2 m_1}\right) + O(T^{\varepsilon - 1})$$

with the error $O(T^{\varepsilon - 1})$ which is admissible. In this way our problem reduces to estimating the following quantity

$$S(H, T, L, M) = \int f(t) \sum_{0 < |h| < H} \sum_{m_1, m_2} a_{m_1} \overline{a}_{m_2}$$

$$\sum_{l_3 l_4 m_2 \equiv -h (\mathrm{mod}\ l_2 m_1)} b(l_2) b(l_3) b(l_4) b\left(l_3 \frac{l_4 m_2}{l_2 m_1}\right) e^{ith/l_3 l_4 m_2} .$$

Without missing crucial points in the demonstration of the arguments we restrict ourselves to consider a partial quantity $S^*(H, T, L, M)$ which denotes the contribution of terms with $(l_4 m_2, l_2 m_1) = 1$. Then the congruence in the innermost sum becomes equivalent to $l_3 \equiv -h\, \overline{l_4 m_2} \pmod{l_2 m_1}$ and the summation over l_3 can be executed by Poisson's formula

$$\sum_{l_3} = \sum_{k\in\mathbb{Z}} e\left(hk\frac{\overline{l_4 m_2}}{l_2 m_1}\right)\int b(\lambda l_2 m_1) b(\lambda l_4 m_2) e^{ith/\lambda l_2 l_4 m_1 m_2} e(\lambda k)\, d\lambda .$$

Here the zero term $(k = 0)$ contains h only in the exponential function $\exp(ith/\lambda l_2 l_4 m_1 m_2)$ which merely oscillates. Nevertheless summation over h yields

$$\sum_{0<|h|<H} e^{ihy} = \frac{e^{iHy} - e^{-iHy}}{e^{iy} - 1} + O(1)$$

where $y = t/\lambda l_2 l_4 m_1 m_2 \ll T^{-\varepsilon}$ and $Hy \gg T^{\varepsilon}$ with the effect that integration over t or λ reduces such terms to admissible quantity.

Considering non-zero terms $(k \neq 0)$ we are led to sums of incomplete Kloosterman sums associated with the Hecke group $\Gamma_o(m_1 m_2)$. The results available from Theorem 19 yield upper bounds which suffice to give

<u>Theorem 20</u> (Deshouillers and Iwaniec). For any complex numbers a_m with $m \leq M \leq T^{1/5}$ we have

$$\int_0^T |\zeta(\tfrac{1}{2} + it)|^4 \Big|\sum_{m\leq M} a_m m^{it}\Big|^2 dt \ll T^{1+\varepsilon} \sum_{m\leq M} |a_m|^2 .$$

Somewhat different arguments were applied in [33] and [11]. Integrals of the kind

$$I(T, M) = \int_0^T |\zeta(\tfrac{1}{2} + it)|^2 \Big|\sum_{M<m\leq 2M} a_m m^{it}\Big|^2 dt$$

may be similarly treated, but unless the coefficients a_m are special, nothing non-trivial was established unconditionally. If the R^*-conjecture of Hooley (see [30]) is assumed then we can show that (see [33])

$$I(T, M) \ll T^{1+\varepsilon} \sum_{M < m \leqslant 2M} |a_m|^2$$

provided $M \leqslant T^{4/7}$. Without any conjecture J.-M. Deshouillers and H. Iwaniec [11] proved this statement for $a_m \equiv \Lambda(m)$.

To a worker in the zeta-function theory, the above power mean-value theorems bring estimates for frequencies of large values taken simultaneously by two Dirichlet polynomials. A noteworthy application is a new upper bound for the difference between consecutive primes; namely

$$p_{n+1} - p_n \ll p_n^{\frac{1}{2} + \frac{1}{21}} .$$

This was proved by J. Pintz. The result with slightly bigger exponent 17/31 was proved independently by the speaker.

The other integrals one may try to estimate by means of Kloosterman sums are the following

$$E(T, H) = \int_T^{T+H} |\zeta(\tfrac{1}{2} + it)|^4 dt$$

where $H = T^\theta$ with $0 < \theta < 1$. In 1979 R. Heath-Brown [25] used A. Weil's estimate for individual Kloosterman sums to evaluate $E(0, T)$ from which it follows that

$$E(T, T^\theta) \ll T^{\theta+\varepsilon}$$

with $\theta = 7/8$. Then the speaker [34] applied the Kuznetsov sum formula and the large sieve inequality for Fourier coefficients of cusp forms to prove this with $\theta = 2/3$. It is interesting to mention that the involved sums of Kloosterman sums are counted with highly oscillating factors. Consequently, after applying Kuznetsov's formula an averaging over the spectrum of the Laplace operator is very long. Thus the circumstances for applying large sieve are not trivial. Unfortunately the sieve is not used in its full force because an averaging over coefficients is short relative to the spectrum. In order to reach a balance we formed another averaging over H-spaces points $T < T_1 < T_2 < \dots < T_R < 2T$, $T_{r+1} - T_r \geqslant H$. Consequently the hybrid large sieve inequality (Theorem 12) is applicable optimally giving

Theorem 21. Let $T^{\frac{1}{2}} < H < T$, $T < T_1 < \dots < T_R < 2T$, $T_{r+1} - T_r \geqslant H$. We then have

$$\sum_{r=1}^{R} E(T_r, H) \ll T^{\varepsilon}(RH + R^{\frac{1}{2}}H^{-\frac{1}{2}}T) .$$

Anyone being familiar with the theory of large values of Dirichlet's polynomials will notice that this theorem contains the well-known result of R. Heath-Brown [26] on the twelfth power-moment

$$\int_0^T |\zeta(\tfrac{1}{2} + it)|^{12}\, dt \ll T^{2+\varepsilon} ,$$

whose original proof does not depend on Kloosterman sums.

Finally let me close this lecture by expressing some prospects in the field. First it seems to be realistic to prove by means now available that

$$S(T, M) \ll T^{\varepsilon}(T + T^{\frac{1}{2}}M^{\frac{3}{2}}) \sum_{m \leqslant M} |a_m|^2 .$$

This result, although too weak to give the correct estimate for the sixth moment of $|\zeta(\frac{1}{2} + it)|$, is powerful enough to yield the density theorem

$$N(\sigma, T) \ll T^{\frac{16}{7}(1-\sigma)+\varepsilon}$$

which is as sharp as the one obtainable by the six moment itself.

The second prospective work refers to power mean-values over Dirichlet's characters

$$\sum_{\substack{\chi \neq \chi_0 \\ \chi (\mathrm{mod}\ q)}} \Big| \sum_{\ell \leq L} \chi(\ell)\ell^{-\frac{1}{2}} \Big|^4 \Big| \sum_{m \leq M} a_m \chi(m) \Big|^2 .$$

Our aim is to estimate this sum from above by

$$O(q^{1+\varepsilon} \sum_{m \leq M} |a_m|^2)$$

with M as large as possible. We can establish it for $M = q^{1/10}$ by using A. Weil's estimate for Kloosterman sums and for $M = q^{\frac{1}{4}}$ by an appeal to Kuznetsov's formula and two conjectures about cusp forms for congruence groups; namely Selberg's eigenvalue conjecture and the Ramanujan - Petersson conjecture. Such a result would contain the following bound of D. Burgess for Dirichlet's L - series

$$L(\tfrac{1}{2} + it, \chi) \ll q^{\frac{3}{16}+\varepsilon} .$$

12. Brun - Titchmarsh Theorem

In 1930 E.C. Titchmarsh [54] used Brun's combinatorial sieve to prove that

$$\pi(x; q, q) \ll \frac{x}{\varphi(q) \log x}$$

uniformly for all $q < x^{1-\varepsilon}$, the constant implied in $\ll$ depending on ε alone. This result, although much less precise than the asymptotic formula

$$\pi(x\,;\,q\,,\,a) \sim \frac{x}{\varphi(q)\ \log x} \qquad \text{as} \quad x \to \infty$$

is very important because it holds for large q. No methods are known to yield an asymptotic formula with $q > x^{\varepsilon}$. The sharpest upper bound obtained by classical sieve methods is

$$\pi(x\,;\,q\,,\,a) \leqslant \frac{2x}{\varphi(q)\ \log \frac{x}{q}}\ ,\ (q\,,\,a) = 1\,,\quad 1 \leqslant q < x$$

due to H.L. Montgomery and R.C. Vaughan [46]. In 1974 Y. Motohashi [47] introduced the pioneering innovation of treating the remainder term in Selberg's sieve nontrivially by means of a large sieve for Dirichlet's characters, thus getting an improvement for $q < x^{2/5}$. Then the speaker, inspired by the works of Y. Motohashi [47] and C. Hooley [28], [29], combined a new version of the combinatorial sieve with estimates for Kloosterman sums to get an improvement in the range $x^{\frac{1}{2}} < q < x^{2/3}$, see [36]. We should mention that it was Hooley who first treated remainder terms in sieve bounds nontrivially and who first applied A. Weil's estimates for Kloosterman sums to the Brun - Titchmarsh theorem. For a survey of these topics, see [36].

Hooley, motivated by applications, studied the problem from the statistical point of view. This involves averaging over q which Hooley [28] and the speaker [36] turned into profit in various ways. Recently J. - M. Deshouillers and the speaker [12] applied Theorem 19 to get

<u>Theorem 22</u> (Deshouillers and Iwaniec). Let $x^{\frac{1}{2}} < Q < x^{1-\varepsilon}$, $a \neq 0$ and $x > x_0(\varepsilon\,,\,a)$. Then

$$\pi(x; q, a) < \frac{(4/3 + \varepsilon_1)x}{\varphi(q) \log \frac{x}{q}},$$

save for at most $Q^{1-\varepsilon_2}$ moduli q in $[Q, 2Q]$.

In this lecture we intend to show briefly how the sums of Kloosterman sums occur in the proof.

We begin by applying the linear sieve of [32] for the sequence $\mathcal{A} = \{l ; l \leqslant x, l \equiv a \pmod q\}$ getting

$$\pi(x; q, a) \leqslant \frac{(2 + \varepsilon)x}{\varphi(q) \log D} + R(D, x, q, a)$$

where parameter D is at our disposal subject to the requirement that $R(D, x, q, a) = O(x/\varphi(q) \log^2 x)$. The total remainder term $R(D, x, q, a)$ consists of a number of bilinear forms

$$B_q(M, N) = \sum_{M < m \leqslant 2M} \sum_{\substack{N < n \leqslant 2N \\ (mn, q) = 1}} a_m b_n r(\mathcal{A}_{mn}; q, a)$$

where $|a_m| \leqslant 1$, $|b_n| \leqslant 1$, $M \leqslant M_o$, $N \leqslant N_o$ with M_o, N_o subject to $M_o N_o = D$. Here the individual error terms are defined by

$$r(\mathcal{A}_d; q, a) = \#\{l \leqslant x, \ l \equiv a \pmod q, \ l \equiv 0 \pmod d\} - \frac{x}{qd}.$$

Letting $\|\xi\| = \xi - [\xi] - \frac{1}{2}$ we get

$$r(\mathcal{A}_d; q, a) = \left\| \frac{-a\bar{d}}{q} \right\| - \left\| \frac{x - a\bar{d}}{q} \right\|.$$

Then after expanding $\|\xi\|$ into Fourier series we arrive at the exponential sums

$$S_q(H, M, N) = \sum_{\substack{M < m \leqslant 2M \\ (mn, q) = 1}} a_m \sum_{N < n \leqslant 2N} b_n \sum_{H < h \leqslant H_1} e\left(ah \frac{\overline{mn}}{q}\right)$$

to be estimated by $O(MNx^{-\varepsilon})$ for any H, H_1 with $H < H_1 \leqslant 2H$, $1 \leqslant H \leqslant MNQx^{-1+\varepsilon}$. Now we perform an averaging over q in $(Q, 2Q]$ and apply Cauchy's inequality to get

$$\sum_{\substack{Q < q \leqslant 2Q \\ (q, a) = 1}} |S_q(H, M, N)| \leqslant \sum_q \sum_m \left| \sum_n b_n \sum_h e\left(ah \frac{\overline{mn}}{q}\right) \right|$$

$$\ll \left\{ QM \sum_q \sum_m \sum_{n_1, n_2} b_{n_1} b_{n_2} \sum_{h_1, h_2} e\left(a(h_1 n_2 - h_2 n_1) \frac{\overline{n_1 n_2 m}}{1}\right) \right\}^{\frac{1}{2}} .$$

The diagonal terms, i.e., such that $h_1 n_2 = h_2 n_1$, contribute trivially $\ll QM(HN)^{\frac{1}{2}} x^{\varepsilon}$, which is admissible provided

$$M < \frac{x}{q} x^{-\varepsilon} .$$

The remaining terms involve sums of incomplete Kloosterman sums associated with the cusps $0, \infty$ of the Hecke group $\Gamma_o(n_1 n_2)$. Applying Theorem 19 we infer an upper bound which is admissible provided

$$N < \left(\frac{x}{q}\right)^{\frac{1}{2}} x^{-\varepsilon} .$$

This together with the previous condition allows us to take the "distribution level" D as large as $\left(\frac{x}{q}\right)^{3/2} x^{-2\varepsilon}$ for almost all q in $(Q, 2Q]$ thereby proving Theorem 22.

It is worthwhile to remark that with the development of modern techniques a new aspect of Brun-Titchmarsh theorem appears, namely that concerning uniformity of results with respect to a residue class $a \pmod q$. Our method depends on Fourier coefficients $\rho_j(a)$ of Maass cusp forms, so unless the Ramanujan-Petersson conjecture that $|\rho_j(a)| \ll a^{\varepsilon}$ is assumed, the resulting bounds depend on the size of a.

Finally I would like to mention the most recent achievement of E. Fouvry [16], who succeeded in evaluating some terms from the combinatorial sieve identities which in the context of the Brun - Titchmarsh theorem were systematically ignored because of a lack of efficient methods.

Theorem 23 (Fouvry). Let $Q = x^{\theta}$, $\frac{1}{2} < \theta < \frac{11}{20}$, $(a, q) = 1$. We then have

$$\pi(x; q; a) \leqslant (c(\theta) - c_1(\theta) + \varepsilon) \frac{x}{\varphi(q) \log x}$$

for almost all q in $[Q, 2Q]$ with the number of exceptions $\ll Q(\log Q)^{-A}$ provided $x > x_o(a, A, \varepsilon)$. Here

$$c(\theta) = \begin{cases} \dfrac{12}{25 - 40\theta} & \text{if } \frac{1}{2} \leqslant \theta \leqslant 53/104 \\[2ex] \dfrac{48}{47 - 56\theta} & \text{if } 53/104 \leqslant \theta \leqslant 11/20 \end{cases}$$

$$c_1(\theta) = \begin{cases} \log \dfrac{10(1-\theta)}{9} & \text{if } \frac{1}{2} \leqslant \theta \leqslant 10/19 \\[2ex] 0 & \text{if } 10/19 \leqslant \theta \leqslant 11/20 . \end{cases}$$

Notice that $c(\frac{1}{2}) - c_1(\frac{1}{2}) = 2.2946\ldots$.

13. Mean-Value Theorems for Arithmetic Progressions

In this lecture I would like to give a brief survey of mean-value theorems of type which are very well exemplified by the celebrated results of E. Bombieri [2] and A.I. Vinogradov [57].

Theorem 24 (Bombieri 1965). For any $A > 0$ there exists $B = B(A) > 0$ such that

$$(1) \qquad \sum_{q \leqslant Q} \max_{(a,q)=1} \max_{y \leqslant x} \left| \pi(y;q;a) - \frac{1}{\varphi(q)} \operatorname{li} x \right| \ll x(\log x)^{-A}$$

with $Q = x^{\frac{1}{2}}(\log x)^{-B}$, where the constant implied in $\ll$ depends on A alone.

I will pay special attention to the works of E. Fouvry and the speaker [14], [15], [17], [18].

1. Auxiliary remarks. The prime number theorem for arithmetic progressions shows that in any residue class a modulo q prime to a there exists asymptotically the same proportion of primes, i.e.,

$$\pi(x; q, a) \sim \frac{\pi(x)}{\varphi(q)} \tag{2}$$

as $x \to \infty$ whenever $(a, q) = 1$, $q \geqslant 1$. Very important is the question of uniformity in q which is closely related to distribution of zeros of Dirichlet's L- series. The well-known theorem of Siegel led Walfisz [58] to the following formula

$$\pi(x; q, a) = \frac{1}{\varphi(q)} \operatorname{li} x + O(x \exp(-c\sqrt{\log x})) \tag{3}$$

for any $q < (\log x)^A$ where c and the constant implied in O depend on A alone (not effectively computable if $A \geqslant 2$). This result, valuable for many problems, is still very far from the truth. Mention should be made of the two conjectures

$$\pi(x; q, a) = \frac{1}{\varphi(q)} \operatorname{li} x + O(x^{\frac{1}{2}+\varepsilon}), \quad \text{Great Riemann Hypothesis}$$

$$\pi(x; q, a) = \frac{1}{\varphi(q)} \operatorname{li} x + O(q^{-\frac{1}{2}} x^{\frac{1}{2}+\varepsilon}), \text{H.L. Montgomery Hypothesis}$$

the first one giving an asymptotic formula valid uniformly for $q < x^{\frac{1}{2} - 2\varepsilon}$ and the latter for $q < x^{1 - 2\varepsilon}$ (cf. [45]). Neither of these conjectures is expected to be proved in the near future.

2. Large sieve approach. With the development of Brun's and Selberg's sieve methods it became popular to investigate formulae which hold for "almost all" moduli q in wider ranges. This became possible after the pioneering work of Yu.V. Linnik [41] (1941) on the large sieve.

Let us give Bombieri's version of the large sieve inequality for Dirichlet's characters:

$$(4) \qquad \sum_{q \leqslant Q} \ \sideset{}{^*}\sum_{\chi (\mathrm{mod}\ q)} \Big| \sum_{n \leqslant N} a_n \chi(n) \Big|^2 \ll (Q^2 + N) \sum_{n \leqslant N} |a_n|^2 .$$

The original proof of Bombieri's mean-value theorem has been greatly simplified by P.X. Gallagher [19] and later R.C. Vaughan [55] offered an "elementary" and still simpler proof. Many generalizations have been established in the meantime. The most remarkable one refers to the so-called Motohashi principle, which, roughly speaking says (we omit minor assumptions):

If two sequences satisfy the assertions of Siegel - Walfisz theorem and Bombieri's mean-value theorem, so does their Dirichlet convolution.

In view of this it is conceivable that the following statements

$$(5) \qquad \sum_{q \leqslant x^{\theta - \varepsilon}} \Big| \sum_{\substack{n \leqslant x \\ n \equiv a (\mathrm{mod}\, q)}} a_n - \frac{1}{\varphi(q)} \sum_{\substack{n \leqslant x \\ (n,q)=1}} a_n \Big| \ll x(\log x)^{-A}$$

are likely to be true for fairly general sequences $\{a_n\}$ with $|a_n| \ll c^{\Omega(n)}$ as long as $\theta \leqslant \frac{1}{2}$. For short, we call θ the level of distribution in arithmetic progressions.

There are essentially three crucial results which are used in all proofs of mean-value theorems for arithmetic progressions:

(i) The Siegel-Walfisz type theorem for small moduli;

$$(6)\qquad \sum_{\substack{n\leqslant x\\ n\equiv a(\mathrm{mod}\ q),(n,f)=1}} a_n = \frac{1}{\varphi(q)}\sum_{\substack{n\leqslant x\\ (n,qf)=1}} a_n + O(\tau^A(f)x(\log x)^{-A}) .$$

(ii) Representation of Σa_n as a sum of bilinear forms

$$(7)\qquad B(M, N: q, a) = \sum_{\substack{m\sim M\\ mn\equiv a(\mathrm{mod}\ q)}} \alpha_m \sum_{n\sim N} \beta_n$$

with $M = x^{\vartheta}$, $N = x^{1-\vartheta}$, $0 < \vartheta < 1$, $|\alpha_m| \leqslant c_1^{\Omega(m)}$, $|\beta_n| \leqslant c_2^{\Omega(n)}$.

(iii) Large sieve inequality (4).

In the case of $a_n = \Lambda(n)$ step (ii) is provided by the Vaughan identity [55].

Let us prove the mean-value theorem for

$$\mathcal{E}(Q, M, N) = \sum_{q\sim Q,\ (q,a)=1} \left|B(M, N; q, a) - \frac{1}{\varphi(q)} B_q(M, N)\right|$$

where $B_q(M, N) = \sum_{(mn,q)=1} \alpha_m\beta_n$ is the expected main term. We advance as follows

$$\mathcal{E}(Q, M, N) \ll \sum_q \frac{1}{\varphi(q)} \sum_{\substack{\chi(\mathrm{mod}\ q)\\ \chi\neq\chi_o}} \left|\sum_m \alpha_m\chi(m)\right| \left|\sum_n \beta_n\chi(n)\right| .$$

Replace each $\chi(\mathrm{mod}\ q)$ with the induced primitive character $\psi(\mathrm{mod}\ e)$, with $q = ef$, $e > 1$ obtaining the upper bound

$$Q^{-1}(\log x)^{c_1} \sum_{f\sim F} \sum_{e\sim E} \sum_{\psi(\mathrm{mod}\ e)}^{*} \left|\sum_{(m,f)=1} \alpha_m\psi(m)\right| \left|\sum_{(n,f)=1} \beta_n\psi(n)\right|$$

with some $EF = Q$. If $E \leqslant E_o = (\log x)^B$ apply (i) giving

$$\mathcal{E}(Q, M, N) \ll x(\log x)^{-A}$$

and if $E_o < E \leqslant Q$ apply Cauchy's inequality and the large sieve inequality (4) giving

$$\mathcal{E}(Q, M, N) \ll \frac{1}{E}(E^2 + M)^{\frac12}(E^2 + N)^{\frac12}(MN)^{\frac12}(\log x)^{c_2}$$

$$\ll (Qx^{\frac12} + MN^{\frac12} + M^{\frac12}N + E_0^{-1}x)(\log x)^{c_2}$$

$$\ll x(\log x)^{-A}$$

provided $Q < x^{\frac12}(\log x)^{-B}$, $M, N < x(\log x)^{-c}$. Hence it is evident that the large sieve inequality sets the limit for the level of distribution, namely $\theta = \frac12$, no matter what value of the 'characteristic number' $\vartheta = \log M/\log x$, of bilinear form $B(M, N; q, a)$ is, provided

$$0 < \vartheta < 1 .$$

In view of this, it is important to realize what can be proved for bilinear forms $B(M, N; q, a)$ if the characteristic number ϑ lies in a specific interval.

3. Dispersion method: First variant. The first successful method was developed in 1980 by E. Fouvry and the speaker [17], [18]. Our basic idea replaces the large sieve by a dispersion method and uses Fourier series technique. Let us sketch the simplest variant.

By Cauchy's inequality the problem of bounding $\mathcal{E}(Q, M, N)$ is reduced to that of bounding the dispersion

$$(8)\quad D(Q, M, N) := \sum_{q \sim Q} \sum_{m \sim M} \Big(\sum_{\substack{n \sim N \\ mn \equiv a[q]}} \beta_n - \frac{1}{\varphi(q)} \sum_{\substack{n \sim N \\ (n,q)=1}} \beta_n \Big)^2 .$$

Needless to say, a non-trivial estimate for the dispersion $D(Q, M, N)$ yields immediately a non-trivial estimate for the error term $\mathcal{E}(Q, M, N)$; a saving of any power of $\log x$ is needed. Before we proceed further, it is already possible to anticipate the condition

$$Q < x^{-\varepsilon}N \qquad (\text{i.e., } \theta < \vartheta) \tag{9}$$

which must be imposed in order to get enough terms β_n to produce cancellation.

Squaring out we write

$$D(Q, M, N) = W(Q, M, N) - 2V(Q, M, N) - U(Q, M, N) \tag{10}$$

with the aim of evaluating each term separately. The most difficult one is

$$W(Q, M, N) = \sum_{q \sim Q} \sum_{m \sim M} \sum_{\substack{n_1, n_2 \sim N \\ mn_1 \equiv mn_2 \equiv a[q]}} \beta_{n_1} \beta_{n_2} .$$

The diagonal terms $n_1 = n_2$ contribute a negligible amount due to our assumption (9). Concerning the terms $n_1 \neq n_2$ we can assume, after a little rearrangement, that they are co-prime. Then the congruences imply

$$n_1 \equiv n_2 \pmod q \tag{11}$$

and

$$m \equiv a\bar{n}_1 \pmod q .$$

We carry out the summation over m first getting

$$\sum_{\substack{m \sim M \\ m \equiv a\bar{n}_1 [q]}} 1 = \frac{M}{q} + r(q, a\bar{n}_1) . \tag{12}$$

For the error terms we trivially get $r(q, a\bar{n}_1) \ll 1$, which obviously is too weak. Therefore we expand $r(q, a\bar{n}_1)$ into a Fourier series that behaves much like

$$\frac{1}{H} \sum_{h \sim H_1} e(ah \frac{\bar{n}_1}{q}) \tag{13}$$

with $H_1 \leqslant H = Q/M$. Here the sum over h alone is too short to yield enough cancellations; indeed if all but one terms cancel out we just miss making errors $r(q, a\bar{n}_1)$ smaller than the main term $q^{-1}M$ in (12). So we look for extra variables n_1, q to average (13). Unfortunately this is impossible to do in a straightforward way, because the congruence (11) constrains n_1 and q too much. For this reason we reinterpret (11) by writing

$$n_1 - n_2 = qr$$

with $0 < |r| < N/Q$, $(r, n_1 n_2) = 1$. Here $|r|$ is rather small and in addition to that we have

$$\frac{\bar{n}_1}{q} \equiv - \frac{\bar{q}}{n_1} + \frac{1}{qn_1} \equiv r\,\frac{\bar{n}_2}{n_1} + \frac{1}{qn_1} \pmod 1 .$$

Since the component $\frac{1}{qn_1}$ can be removed by partial summation, we are led to sums of the type

$$\frac{1}{H} \sum_h \sum_r \sum_{\substack{(n_1, n_2) = 1 \\ n_1 \equiv n_2 [r]}} \beta_{n_1} \beta_{n_2}\, e(ahr\,\frac{\bar{n}_2}{n_1}) .$$

After applying Cauchy's inequality we get

$$\Big| \sum_{n_1, n_2} \beta_{n_1} \beta_{n_2}\, e(ahr\,\frac{\bar{n}_2}{n_1}) \Big|^2 \ll N \sum_{n_1, n_1'} \Big| \sum_{n_2} e(ahr(n_1' - n_1)\,\frac{\bar{n}_2}{n_1' n_1}) \Big| .$$

The innermost sum is an incomplete Kloosterman sum; its length is about N while the modulus is $n_1 n_2 \sim N^2$. Then an easy calculation shows that A. Weil's result yields only the trivial bound. The Hooley conjecture R^* [30] would be useful. Yet, it is possible to avoid any unproved hypothesis if one knows, for example, that β is the convolution of two sequences $\beta = \lambda * \delta$, each one with essentially shorter support than that of β. In such circumstances Cauchy's inequality can be applied in a different manner so that the resulting incomplete Kloosterman sums are longer and Weil's estimate is useful.

The other terms $V(Q, M, N)$ and $U(Q, M, N)$ from the dispersion can be treated in a similar way and in practice they are much easier. As regards main terms we obtain the following square mean-value:

$$M \sum_q (\varphi(q))^{-2} \sum^*_{\ell\,(\mathrm{mod}\,q)} \Big| \sum_{n \equiv \ell\,(\mathrm{mod}\,q)} \beta_n - \frac{1}{\varphi(q)} \sum_{(n,q)=1} \beta_n \Big|^2 .$$

In order to estimate this we apply a generalized form of the Barban - Davenport - Halberstam theorem.

Concluding the above investigations let us say that the method yields non-trivial bound for the dispersion $D(Q, M, N)$ if the characteristic number ϑ and the level of distribution θ satisfy

$$\theta < \vartheta < \frac{4}{7}\,, \tag{14}$$

provided Hooley's R^* conjecture is true.

4. <u>Dispersion method: Second variant (dual)</u>. The total error term $\mathcal{E}(Q, M, N)$ is written as

$$\mathcal{E}(Q, M, N) = \sum_{\substack{q \sim Q \\ (q,a)=1}} \varepsilon_q \Big(\sum_{m \sim M} \sum_{\substack{n \sim N \\ mn \equiv a[q]}} \alpha_m \beta_m - \frac{1}{\varphi(q)} \sum_{m \sim M} \sum_{\substack{n \sim N \\ (mn,q)=1}} \alpha_m \beta_n \Big)$$

where $\varepsilon_q = \pm 1$, whence

$$\mathcal{E}(Q, M, N) \ll \sum_m \Big| \sum_{\substack{(q, m)=1 \\ q \sim Q}} \varepsilon_q \Big(\sum_{\substack{n \sim N \\ mn \equiv a[q]}} \beta_n - \frac{1}{\varphi(q)} \sum_{\substack{n \sim N \\ (n,q)=1}} \beta_n \Big) \Big| .$$

Then by the Cauchy - Schwarz inequality the problem reduces to bounding the dispersion

$$D(Q, M, N) = \sum_m \Big|\sum_q \varepsilon_q \Big(\sum_n \beta_n - \frac{1}{\varphi(q)} \sum_n \beta_n \Big) \Big|^2$$

$$= W(Q, M, N) - 2V(Q, M, N) + U(Q, M, N) .$$

The most difficult term $W(Q, M, N)$ is equal to

$$W(Q, M, N) = \sum_{q_1, q_2} \varepsilon_{q_1} \varepsilon_{q_2} \sum_{n_1, n_2} \beta_{n_1} \beta_{n_2} \sum_{\substack{m \equiv a\bar{n}_1 [q_1] \\ m \equiv a\bar{n}_2 [q_2]}} 1 .$$

Using the Fourier series technique and estimates for Kloosterman sums (Theorem 19) Fouvry [14], [15] was able to estimate the dispersion non-trivially in the following range

$$\theta < \min \Big(\frac{1 + \vartheta}{2} , \frac{5 - 6\vartheta}{8} \Big) \tag{15}$$

which admits $\theta > \frac{1}{2}$ for $0 < \vartheta < \frac{1}{6}$.

5. Some applications. Of many applications of (15) let us mention the following:

Theorem 25 (Fouvry 1982). Let $z = x^{\vartheta}$, $0 < \vartheta < \frac{1}{6}$ and $f_z(n)$ be the characteristic function of numbers free of prime factors $\leqslant z$. Then the level of distribution of $f_z(n)$ is $\theta = \theta(\vartheta) > \frac{1}{2}$.

Theorem 26 (Fouvry 1982). If Bombieri's mean-value theorem holds for the divisor functions $\tau_k(n)$, $k \leqslant 6$ (with certain uniformity) with the level of distribution $\theta > \frac{1}{2}$, so it does for $\Lambda(n)$.

6. Mean-value theorem for primes. So far we were able to fill up rather little of the interval $0 < \vartheta < \frac{1}{2}$. It is unfortunate that for getting a mean-value theorem with $\theta > \frac{1}{2}$ for primes one has to deal with bilinear forms of arbitrary characteristic ϑ from $0 < \vartheta \leqslant \frac{1}{2}$. To overcome this difficulty we began (jointly with Fouvry) to study forms of the type

$$\mathcal{E}(Q, R; M, N) = \sum_{q\sim Q} \alpha_q \sum_{r\sim R} \beta_r \Big(\sum_{\substack{m\sim M \\ mn \equiv a[qr]}} \gamma_m \sum_{n\sim N} \delta_n - \frac{1}{\varphi(qr)} \sum_{(mn,qr)=1} \gamma_m \delta_n \Big)$$

$$\ll \sum_q \sum_m \Big| \sum_r \beta_r \Big(\sum_n \delta_n - \frac{1}{\varphi(qr)} \sum_n \delta_n \Big|$$

and the corresponding dispersions. This builds up greater flexibility into arguments described in the two previous approaches. And consequently we were able to extend the results as follows:

Let $N = x^{\vartheta}$, $M = x^{1-\vartheta}$, $Q = x^{\theta_1}$, $R = x^{\theta_2}$, $\theta_1 + \theta_2 = \theta$.

Then, omitting minor assumptions, we have

$$\mathcal{E}(Q, R, M, N) \ll x(\log x)^{-A}$$

provided one of the following situations takes place:

(a) δ_n - satisfies the Siegel-Walfisz theorem, and

$$0 < \vartheta < \frac{1}{3}, \quad \theta_1 < \vartheta, \quad \theta_2 < \min\Big(\frac{2-3\vartheta}{4}, \frac{5-11\vartheta}{8}\Big),$$

(b) $\delta_n \equiv 1$, and

$$\tfrac{1}{4} < \vartheta \leqslant 1, \quad \theta_1 < \vartheta, \quad \theta_2 < \frac{1-\vartheta}{3}.$$

Observe that in either case the constraints permit taking $\theta = \theta_1 + \theta_2 > \frac{1}{2}$. Applying a combinatorial identity of Linnik [42] or Heath-Brown [27] for the Mongoldt function $\Lambda(n)$ we always may fall into one of the two situations stated. The only remaining problem is that the modulus qr of the arithmetic progressions must be factorized appropriately for the change of the characteristic ϑ of bilinear forms that need be considered.

To make use of the above results we introduce the following concept of a <u>well-factorable</u> function. We say that $\lambda(q)$ is well-factorable of level Q if for any $Q_1 Q_2 = Q$ there are two functions $\lambda_1(q_1)$, $\lambda_2(q_2)$ supported in $(0, Q_1]$ and $(0, Q_2]$ respectively and bounded by 1 in absolute value such that

$$\lambda = \lambda_1 * \lambda_2 .$$

From what we said follows

<u>Theorem 27</u> (Fouvry and Iwaniec 1982). Let $a \neq 0$, $A > 0$, $\varepsilon > 0$, $x \geqslant 2$. For any well-factorable function $\lambda(q)$ of level $Q = x^{9/17 - \varepsilon}$ we have

$$\sum_{(q, a) = 1} \lambda(q)\left(\pi(x; q, a) - \frac{1}{\varphi(q)} \operatorname{li} x\right) \ll x(\log x)^{-A},$$

the constant implied in $\ll$ depending at most on ε, a, A.

Perhaps we should give a better motivation for introducing well-factorable functions: we wish to point out that they occur as weight-functions in the remainder terms of the linear sieve, cf. [32].

Finally it should be noted that our method does not allow the residue class a to vary with q. Moreover, a cannot be too large. The main reason is the lack of good estimates for Fourier coefficients $\rho_j(a)$ of Maass cusp forms which are implicitly employed here by an appeal to Theorem 19. The Ramanujan - Petersson conjecture that

$$|\rho_j(a)| \ll |a|^{\varepsilon}$$

would perhaps permit one to take $|a|$ as large as x.

14. An Additive Divisor Problem

In this last lecture we discuss a problem of evaluating the sums

$$D(x, h) = \sum_{|h| < n \leqslant x} d(n)d(n + h) .$$

A relation of these sums with automorphic functions is not new. The divisor function $d(n)$ proves to be the nth Fourier coefficient of the Eisenstein series $E(z, s)$ at $s = \frac{1}{2}$. Thus the generating Dirichlet's series

$$R(s) = \Sigma\, d(n)d(n + h)(n + \frac{h}{2})^{-s}$$

is a kind of Rankin-Selberg zeta-function. Yet, A. Selberg [52] gave a hint as to how the analytic nature of such zeta-functions can be established by means of the spectral theory of $L^2(\Gamma \backslash H)$. For a detailed analysis see N.V. Kuznetsov [40]. It turns out that the main term for $D(x, h)$ is $xP_h(\log x)$, where $P_h(X)$ is a quadratic polynomial with the leading coefficient $6\pi^{-2}\mathfrak{S}_{-1}(h)$. Several mathematicians established upper bounds for the error term

(1) $$E(x, h) := D(x, h) - xP_h(\log x) \ll x^{\theta + \varepsilon}$$

with $\theta = \frac{11}{12}$ (Estermann [13]) and $\theta = 5/6$ (Heath-Brown [25] and Wirsing -- unpublished). They all used estimates for Kloosterman sums. That Kloosterman sums do something for this problem is evident when one tries to solve the relevant equation

(2) $$ad - bc = h \quad \text{with} \quad |h| < ad \leqslant x$$

by the "ad hoc" method. Interpreting (2) as a congruence $a \equiv h\bar{d} \pmod{c}$ one counts the solutions of the latter by using the Fourier series technique, and the error term is controlled by a sum of incomplete Kloosterman sums. Following this line J.-M. Deshouillers and H. Iwaniec [9] applied the Kuznetsov sum formula (Theorem 9) and the large sieve inequality (Theorem 11) to prove (1) with $\theta = 2/3$. This result seems to be the best possible for the present methods. If, however, an additional averaging over h is performed we can get a bit stronger inequality

$$\sum_{1 \leqslant h \leqslant H} |E(x, h)| \ll x^{\frac{1}{2} + \varepsilon} H + x^{\frac{2}{3} + \varepsilon} H^{\frac{1}{2}}$$

for $1 \leqslant H \leqslant x^{\frac{1}{2}}$, where the constant implied in $\ll$ depends on ε alone.

We wish to call the reader's attention to the short note of D. Hejhal [24] in which the author announced an asymptotic formula with the error term $O(x^{2/3 + \varepsilon})$ for the mean-value of another arithmetic function similar to $d(n)d(n + h)$.

Concerning estimates of allied sums

$$\sum_{n \leqslant x} c(n)c(n + h)$$

with Fourier coefficients $c(n)$ of cusp forms the reader is referred to papers of D. Goldfeld [20], J. Hafner [23] and A. Good [22].

J.-M. Deshouillers pointed out that the estimates for sums of Kloosterman sums associated with congruence groups (Theorem 19) can be used to establish an asymptotic formula for

$$\sum_{|h| < n \leqslant x} d_3(n)d(n+h)$$

with the error term $O(x^{1-\eta})$, $0 < \eta < 1$. The allied sum

$$\sum_{|h| < n \leqslant x} d_3(n)d_3(n+h)$$

has not yet been evaluated asymptotically. An attempt of L.A. Takhtajan and A.I. Vinogradov [53], doubtless interesting, did not reach the aim.

ACKNOWLEDGEMENTS

I would like to express my thanks to The University of Texas for financial support and to Professor J. Vaaler and his wife for offering me their exceptional help. A very nice scientific atmosphere was created by the organizers and the participants. In my lectures I have benefitted from conversations with several mathematicians, first of all with Professors P. Sarnak, D. Goldfeld, J. Hafner, I. Vardi, J. Friedlander and A. Ghosh. Several results discussed here have been established by Professor J.-M. Deshouillers jointly with the speaker. I am glad to thank Prof. Deshouillers for the cooperation. I am grateful to Prof. A. Schinzel for reading this text and indicating some corrections.

Warsaw, December 1982

REFERENCES

[1] Atkin, A.O.L. and J. Lehner, "Hecke operators on $\Gamma_0(m)$," Math. Ann. 185 (1970), 134-160.

[2] Bombieri, E., "On the large sieve," Mathematika 19 (1965), 201-225.

[3] Bruggeman, R.W., "Fourier coefficients of cusp forms," Invent. Math. 45 (1978), 1-18.

[4] ------, Fourier coefficients of automorphic forms, Lect. Notes in Math., vol. 865, Berlin-New York, Springer 1981.

[5] Burgess, D., "On character sums and L-series, II," Proc. London Math. Soc. (3)13 (1963), 524-536.

[6] Deligne, P., Formes modulaires et représentation l-adiques, Lect. Notes in Math., vol. 179, Berlin-New York, Springer 1971, 139-172.

[7] ------, "La conjecture de Weil I, Publ. Math. I.H.E.S. 43 (1974), 273-307.

[8] Deshouillers, J.-M. and H. Iwaniec, "Kloosterman sums and Fourier coefficients of cusp forms," Invent. Math. 70 (1982), 219-288.

[9] ------, "An additive divisor problem," J. London Math. Soc. (2)26 (1982), 1-14.

[10] ------, "Power mean-values for the Riemann zeta-function," Mathematika, 29 (1982), 202-212.

[11] ------, "Power mean-values for the Riemann zeta-function, II," Acta Arith. 43 (1983), 95-102.

[12] ------, "On the Brun-Titchmarsch theorem and the greatest prime factor of p + a," Proc. Janos Bolyai Soc. Conf. 1981 (to appear).

[13] Esterman, T., "Über die Darstellung einer Zahl als Differenz von zwei Produkten," J. Reine Angew. Math. 164 (1931), 173-182.

[14] Fouvry, E., "Répartition des suites dans les progressions arithmétiques," Acta Arith. 41 (1982), 359-382.

[15] ------, "Autour du théorème de Bombieri-Vinogradov," (to appear).

[16] ------, "Sur le théorème de Brun-Titchmarsch," Acta Arith. 43 (to appear).

[17] Fouvry, E. and H. Iwaniec, "On a theorem of Bombieri-Vinogradov type," Mathematika 27 (1980), 135-152.

[18] ------, "Primes in arithmetic progressions," Acta Arith. 42 (1983), 197-218.

[19] Gallagher, P.X., "Bombieri's mean value theorem," Mathematika 15 (1968), 1-6.

[20] Goldfeld, D., "Analytic and arithmetic theory of Poincaré series," Journées Arithmetiques, Asterisque 61 (1979), 95-107.

[21] ------, and P. Sarnak, "Sums of Kloosterman Sums," Invent. Math. 71 (1983), 243-250.

[22] Good, A., "On various means involving the Fourier coefficients of cusp forms," Math. Zeit. 183 (1983), 95-129.

[23] Hafner, J., "Explicit estimates in the arithmetic theory of cusp forms and Poincaré series," Preprint 1982.

[24] Hejhal, D., "Sur certaines séries de Dirichlet dont les pôles sont sur les lignes critiques," C.R. Acad. Sc. Paris 287, série A (1978), 383-385.

[25] Heath-Brown, D.R., "The fourth power moment of the Riemann zeta-function," Proc. London Math Soc. (3)38 (1979), 385-422.

[26] ------, "The twelfth power moment of the Riemann zeta-function," Quart. J. Math. Oxford (2)29 (1978), 443-462.

[27] ------, "Sieve identities and gaps between primes," Journées Arithmétiques de Metz 1981 (to appear in Asterisque).

[28] Hooley, C., "On the Brun-Titchmarsch theorem," J. Reine Angew. Math. 255 (1972), 60-79.

[29] ------, "On the Brun-Titchmarsch theorem, II," Proc. London Math. Soc. (3)30 (1975), 114-128.

[30] ------, "On the greatest prime factor of a cubic polynomial," J. Reine Angew. Math. 303/304 (1978), 21-50.

[31] Iwaniec, H., "Prime geodesic theorem," J. Reine Angew. Math. 349 (1984), 136-159.

[32] ------, "A new form of the error term in the linear sieve," Acta Arith. 37 (1980), 307-320.

[33] ------, "On mean values for Dirichlet's polynomials and the Riemann zeta-function," J. London Math. Soc. 22 (1980), 39-45.

[34] ------, "Fourier coefficients of cusp forms and the Riemann zeta-function," Sém. Th. Nb. Bordeaux (1979-80), exposé No. 18, 36 pages.

[35] ------, "Mean values for Fourier coefficients of cusp forms and sums of Kloosterman sums," Journée Arithmétiques de Exeter 1980, Cambridge University Press 1982.

[36] ------, "On the Brun-Titchmarsch theorem," J. Math. Soc. Japan (1)34 (1982), 95-122.

[37] ------, and J. Szmidt, "Density theorems for exceptional eigenvalues of Laplacian for congruence groups," to appear in the Banach Center Publications, vol. 17.

[38] Kubota, T., Elementary Theory of Eisenstein Series, New York: John Wiley and Sons, 1973.

[39] Kuznetsov, N.V., "Petersson hypothesis for parabolic forms of weight zero and Linnik hypothesis. Sums of Kloosterman sums," (in Russian), Math. Sbornik 111 (153), No. 3 (1980), 334-383.

[40] ------, "Spectral methods in arithmetical problems," (in Russian), Zap. Naučn. Sem. Leningrad. Otdel. Mat. Inst. Steklov (LOMI) (1978), vol. 76, 159-166.

[41] Linnik, Yu.V., "The large sieve," (in Russian), Dokl. Akad. Nauk USSR 30 (1941), 292-294.

[42] ------, "The dispersion Method in Binary Additive Problems," Translations of Mathematical Monographs 4, AMS Providence, 1963.

[43] ------, "Additive problems and eigenvalues of the modular operators," Proc. Internat. Congr. Math. Stockholm (1962), 270-284.

[44] Montgomery, H.L., Topics in Multiplicative Number Theory, Lect. Notes in Math., vol. 227, Berlin-New York, Springer 1971.

[45] ------, "Problems concerning prime numbers," Proc. Symp. Pure Math. 28 (1976), 307-310.

[46] ------, and R.C. Vaughan, "On the large sieve," Mathematika 20 (1973), 119-134.

[47] Motohashi, Y., "On some improvements of the Brun-Titchmarsch theorem," J. Math. Soc. Japan 26 (1974), 306-323.

[48] Randol, B., "Small eigenvalues of the Laplace operator on compact Riemann surfaces," Bul. AMS 80 (1974), 996-1000.

[49] Rankin, R., "Contributions to the theory of Ramanujan's function $\tau(n)$ and similar arithmetical functions," Proc. Cambridge Philos. Soc. 35 (1939), 357-372.

[50] Rankin, R., "The vanishing of Poincaré series," Proc. Edinburgh Math. Soc. 23 (1980), 151-161.

[51] Selberg, A., "Harmonic analysis and discontinuous groups in symmetric Riemannian spaces with applications to Dirichlet's series," J. Indian Math. Soc. 20 (1956), 41-87.

[52] ------, "On the estimation of Fourier coefficients of modular forms," Proc. Symp. Pure Math. 8 (1965), 1-15.

[53] Takhtajan, L.A. and A.I. Vinogradov, "Theory of Eisenstein series for the group SL(3, ℝ) and its application to a binary problem. Part I, Fourier expansion of the highest Eisenstein series," (in Russian), Zap. Naučn. Sem. Leningrad. Otdel. Mat. Inst. Steklov (LOMI) (1978), vol. 76, 5-52.

[54] Titchmarsch, E.C., "A divisor problem," Rend. Circ. Mat. Palermo 54 (1930), 414-429.

[55] Vaughan, R.C., "An elementary method in prime number theory," Acta Arith. 37 (1980), 111-115.

[56] Venkov, A.B., "Spectral theory of automorphic functions," (in Russian), Trudy Mat. Inst. Sekl., CL III 1981, 170.

[57] Vinogradov, A.I., "The density hypothesis for Dirichlet L-series," (in Russian), Izv. Akad. Nauk USSR Ser. Mat. 29 (1965), 903-934.

[58] Walfisz, A., "Zur additiven Zahlentheorie II," Math. Z. 40 (1936), 592-607.

The Distribution of r-Free Integers in Arithmetic Progressions

Kevin S. McCurley

1. INTRODUCTION.

A natural number is called r-free if it is not divisible by the rth power of a prime. Let $S_r(x;q,a)$ denote the number of r-free numbers in the arithmetic progression a modulo q that do not exceed x, and let

$$R_r(x;q,a) = S_r(x;q,a) - \frac{x}{q} f(a,q) ,$$

where

$$f(a,q) = \sum_{\substack{d=1 \\ (d^r,q)|a}}^{\infty} \frac{\mu(d)(d^r,q)}{d^r} .$$

We shall always assume that (a,q) is r-free, for otherwise $S_r(x;q,a)$ is zero.

In this paper, we shall be concerned with the estimation of $R_r(x;q,a)$, with emphasis on uniformity in a, q, and r. We shall use $c_1, c_2 \ldots$ to denote constants, and unless otherwise indicated all constants will be independent of a, q, and r. Our starting point is the formula

$$(1) \qquad S_r(x;q,a) = \sum_{\substack{n \leqslant x \\ n \equiv a (\mathrm{mod}\ q)}} \sum_{d^r|n} \mu(d)$$

from which an elementary argument yields the estimate

$$(2) \qquad R_r(x;q,a) \ll x^{1/r} .$$

In the case (a,q) = 1 a more elaborate argument due to Prachar [6] yields

$$(3) \qquad R_r(x;q,a) \ll r^{\omega(q)}\{x^{1/r}q^{-1/r^2} + q^{1/r}\} \quad ,$$

where $\omega(q)$ is the number of distinct prime factors of q. For $r = 2$ this has been improved by Hooley [4] to

$$(4) \qquad R_2(x;q,a) \ll x^{1/2}q^{-1/2} + q^{1/2+\epsilon} \quad ,$$

and the recent work of Heath-Brown [3] would seem to imply a stronger result than (4), at least in the case $x \leqslant q^2$.

The previous results are primarily concerned with small values of x relative to q, whereas in this paper we shall be concerned with large values of x. Siebert [7] proved that if $\epsilon > 0$ is arbitrary and $x \geqslant \exp(q^{\epsilon})$, then

$$(5) \qquad R_r(x;q,a) \ll x^{1/r} \exp(-c_1(\epsilon,r)\sqrt{\log x}) \quad .$$

This result is analogous to the Siegel-Walfisz theorem for primes in arithmetic progressions (see Davenport [2], p. 132). The distribution of primes in arithmetic progressions modulo q depends on the location of zeros of Dirichlet L-functions formed with characters modulo q. We say that q is an exceptional modulus if there exists a real character modulo q such that the associated L-function has a real zero exceeding $1 - c_2/\log q$. Page [5] proved that if q is not an exceptional modulus, then the Siegel-Walfisz theorem can be substantially improved.

Our first result is an improvement of (5) that is analogous to Page's theorem.

THEOREM 1. There exist absolute computable constants c_3 and c_4 such that if $x \geqslant \exp(c_3 r \log^2 q)$ and q is not exceptional, then

$$R_r(x;q,a) \ll (xq)^{1/r} \exp(-c_4 r^{-3/2}\sqrt{\log x}) \ .$$

Note that this is inferior to (1) unless $x \geqslant \exp(c_4^{-2} r \log^2 q)$. Theorem 1 may also be regarded as a generalization of a result of Walfisz [8, pp. 192-198], who proved it for $q = 1$.

The proof of Theorem 1 is similar to that of Siebert [7], and is based on an estimate for the functions

$$M(x;q,a) = \sum_{\substack{n \leqslant x \\ n \equiv a \pmod q}} \mu(n) \ .$$

This requires information concerning the zeros of all L-functions formed with characters modulo q. In our next theorem we use a slightly different method to show that we need only be concerned with characters of the form χ^r. In contrast with previous methods, we now assume that $(a,q) = 1$.

THEOREM 2. Let q and r be such that $L(s,\chi^r)$ has no real zeros exceeding $1 - c_2/\log q$, for all χ modulo q. Then there exist constants c_6 and c_7 such that

$$R_r(x;q,a) \ll x^{1/r} \exp(-c_6 r^{-3/2} \sqrt{\log x}) \ ,$$

provided $x \geqslant \exp(c_7 r \log^2 q)$ and $(a,q) = 1$.

From Theorem 2 it is apparent that the problem of exceptional moduli for r-free numbers is different from that of primes. For example, if $r = \varphi(q)$ then χ^r is principal and $L(s,\chi^r)$ has no positive real zeros. For a given r, the moduli that are potentially troublesome for r-free numbers are those for which there exist characters χ with χ^r a quadratic character. In the case of squarefree numbers, this has the following consequence.

Corollary. There exist constants c_8 and c_9 such that if $x \geqslant \exp(c_8 \log^2 q)$, $(a,q) = 1$, $16 \nmid q$, and q is not divisible by a prime congruent to 1 modulo 4, then

$$R_2(x;q,a) \ll x^{1/2} \exp(-c_9 \sqrt{\log x}) .$$

In order to prove the Corollary, we write $q = p_1^{\alpha_1} p_2^{\alpha_2} \cdots p_k^{\alpha_k}$, and write a character χ modulo q as $\chi = \chi_1\chi_2 \ldots \chi_k$, where χ_i is a character modulo $p_i^{\alpha_i}$. The order of χ is then the least common multiple of the order of the characters χ_i, so it suffices to prove that χ_i is not of order 4. If p_i is odd, then the order of χ_i divides $\varphi(p_i^{\alpha_i}) = (p_i - 1)p_i^{\alpha_i - 1}$, and this is not divisible by 4 if $p_i \equiv 3 \pmod 4$. Finally we observe that there are no quartic characters modulo 1, 2, 4, or 8.

2. PRELIMINARIES.

The following result is analogous to a result of Page.

Lemma 1. There exist constants c_{10} and c_{11} such that if $L(s,\chi)$ has no real zeros exceeding $1 - c_2/\log q$, and $x \geqslant \exp(c_{10} \log^2 q)$, then

$$\sum_{n \leqslant x} \mu(n)\chi(n) \ll x \exp(-c_{11} \sqrt{\log x}) .$$

The proof of Lemma 1 is omitted, since it is implicit in the work of Davenport [1]. From Lemma 1 we immediately obtain the following lemmas.

Lemma 2. If $x \geqslant \exp(c_{10} \log^2 q)$ and q is not exceptional, then

$$M(x;q,\ell) \ll x \exp(-c_{11} \sqrt{\log x}) .$$

Proof. If $(\ell, q) = 1$, then

$$M(x;q,\ell) = \frac{1}{\varphi(q)} \sum_{\chi} \bar{\chi}(\ell) \sum_{n \leqslant x} \chi(n)\mu(n) ,$$

and the result follow from Lemma 1. The case $(\ell, q) > 1$ is similarly handled in Davenport [1].

Lemma 3. If $x \geqslant \exp(c_{10}\log^2 q)$, $r \geqslant 2$, and q is not exceptional, then

$$\sum_{\substack{n > x \\ n \equiv \ell \pmod q}} \mu(n)n^{-r} \ll x^{1-r} \exp(-c_{11}\sqrt{\log x}) .$$

Lemma 3 follows from Lemma 2 by partial summation. Similarly if we take χ principal in Lemma 1, then we obtain the following by partial summation.

Lemma 4. If $x \geqslant \exp(c_{10}\log^2 q)$ and $r \geqslant 2$, then

$$\sum_{\substack{n > x \\ (n,q)=1}} \mu(n)n^{-r} \ll x^{1-r} \exp(c_{11}\sqrt{\log x}) .$$

3. PROOF OF THEOREM 1.

The methods used by Walfisz and Siebert bear a resemblance to the "hyperbola method" used in the Dirichlet divisor problem. If $y \leqslant x^{1/r}$, then (1) yields

$$S_r(x;q,a) = \sum_{\substack{d^r m \leqslant x \\ d^r m \equiv a \pmod q}} \mu(d) \tag{6}$$

$$= \sum_{d \leqslant y} + \sum_{d > y}$$

$$= \sum_1 + \sum_2$$

say. Then

$$\sum_1 = \sum_{d \leqslant y} \mu(d) \sum_{\substack{m \leqslant xd^{-r} \\ d^r m \equiv a \pmod q}} 1 ,$$

and clearly

$$\sum_{\substack{m \leqslant xd^{-r} \\ d^r m \equiv a \pmod q}} 1 = \begin{cases} 0 , & (d^r,q) \nmid a , \\ \dfrac{x(d^r,q)}{qd^r} + O(1) , & (d^r,q) \mid a . \end{cases}$$

Hence

$$\sum_1 = \frac{x}{q} f(a,q) - \frac{x}{q} \sum_{\substack{\ell=1 \\ (\ell^r,q) \mid a}}^{q} (\ell^r,q) \sum_{\substack{d > y \\ d \equiv \ell \pmod q}} \frac{\mu(d)}{d^r} + O(y) .$$

If $y \geqslant \exp(c_{10} \log^2 q)$, then Lemma 3 yields

$$\sum_1 = \frac{x}{q} f(a,q) + O(xy^{1-r} \exp(-c_{11} \sqrt{\log y})) + O(y) . \tag{7}$$

From Lemma 2 it follows directly that

$$\sum_2 = \sum_{m \leqslant xy^{-r}} \sum_{\substack{\ell=1 \\ \ell^r m \equiv a \pmod q}}^{q} \{M((x/m)^{1/r};q,\ell) - M(y;q,\ell)\}$$

$$\ll x^{1/r} q \exp(-c_{11} \sqrt{\log y}) \sum_{m \leqslant xy^{-r}} m^{-1/r}$$

$$+ xqy^{1-r} \exp(-c_{11} \sqrt{\log y})$$

$$\leqslant \; xqy^{1-r} \exp(-c_{11}\sqrt{\log y}) \; .$$

We then choose $y = (xq)^{1/r} \exp(-\frac{c_{11}}{\sqrt{2}} r^{-3/2}\sqrt{\log x})$. Note that

$$\log y \geqslant r^{-1} \log x - c_{11} r^{-3/2}\sqrt{\log x}$$

$$\geqslant \frac{1}{2r} \log x$$

$$\geqslant c_{10} \log^2 q \quad ,$$

if x is sufficiently large and $c_3 \geqslant 2c_{10}$. Finally,

$$\log y = \frac{1}{r} \log x + \frac{1}{r} \log q - \frac{c_{11}}{\sqrt{2}} r^{-3/2}\sqrt{\log x}$$

$$\leqslant \frac{1}{r} \log x + \frac{1}{r} \log q - c_{11}\sqrt{\frac{c_3}{2}}\, r^{-1} \log q$$

$$\leqslant \frac{1}{r} \log x$$

if $c_3 \geqslant 2c_{11}^{-2}$, so that $y \leqslant x^{1/r}$.

4. PROOF OF THEOREM 2.

We use (6) again but estimate the sum $\sum_2$ in a different way. If $(a,q) = 1$, then

$$\text{(8)} \qquad \sum\nolimits_2 = \sum_{\substack{md^r \leqslant x \\ md^r \equiv a \,(\mathrm{mod}\ q) \\ d > y}} \mu(d)$$

$$= \frac{1}{\varphi(q)} \sum_{\chi} \bar{\chi}(a) \sum_{\substack{md^r \leqslant x \\ d > y}} \mu(d)\chi^r(d)\chi(m)$$

By Lemma 1 the inner sum satisfies

$$\sum_{m \leqslant xy^{-r}} \chi(m) \left\{ \sum_{d \leqslant (x/m)^{1/r}} \mu(d)\chi^r(d) - \sum_{d \leqslant y} \mu(d)\chi^r(d) \right\}$$

$$\ll x^{1/r} \exp(-c_{11}\sqrt{\log y}) \sum_{m \leqslant xy^{-r}} m^{-1/r} + xy^{1-r} \exp(-c_{11}\sqrt{\log y})$$

$$\ll xy^{1-r} \exp(-c_{11}\sqrt{\log y}) \quad .$$

This time we choose $y = x^{1/r} \exp(-\frac{c_{11}}{\sqrt{2}} r^{-3/2}\sqrt{\log x})$, and the result follows from (6), (7), and (8).

Department of Mathematics
Michigan State University
East Lansing, Michigan 48824

Current Address:
Department of Mathematics
DRB 306, University Park
University of Southern California
Los Angeles, California 90089-1113

REFERENCES

[1] Davenport, H., "On some infinite series involving arithmetical functions (II)", Quarterly J. Math. 8 (1935), 313 - 320.

[2] ------, Multiplicative Number Theory, second edition revised by Hugh Montgomery, Springer-Verlag, New York, (1980).

[3] Heath-Brown, D.R., "The least square-free number in an arithmetic progression", J. Reine Angew. Math. 332 (1982), 204 - 220.

[4] Hooley, C., "A note on square-free numbers in arithmetic progressions", Bull. London Math. Soc. 7(1975), 133-176.

[5] Page, A., "On the number of primes in an arithmetic progression", Proc. London Math. Soc. (2) 39(1935), 116 - 141.

[6] Prachar, K., "Über die kleinste quadratfreie Zahl einer arithmetischen Reihe", Monatsh. Math. 62 (1958), 173 - 176.

[7] Seibert, H., "Einige Analoga zum Satz von Siegel-Walfisz", Zahlentheorie, Ber. Math. Forschungsinst., Oberwolfach, Heft 5, Bibliographisches Inst., Mannheim, 1971, 173 - 184.

[8] Walfisz, A., Weylsche Exponentialsummen in der neueren Zahlentheorie, Berlin, 1963.

Printed and bound by CPI Group (UK) Ltd, Croydon, CR0 4YY

27/10/2024

14580154-0003